MIKE MEYERS' CERTIFICATION

Passport ★

CompTIA®

RFID+™
Certification

MARK BROWN

SAM PATADIA

SANJIV DUA

Mc Graw Hill

New York • Chicago • San Francisco
Lisbon • London • Madrid • Mexico City
Milan • New Delhi • San Juan
Seoul • Singapore • Sydney • Toronto

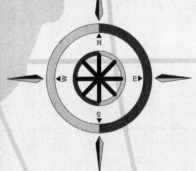

The **McGraw·Hill** Companies

Cataloging-in-Publication Data is on file with the Library of Congress

McGraw-Hill books are available at special quantity discounts to use as premiums and sales promotions, or for use in corporate training programs. For more information, please write to the Director of Special Sales, Professional Publishing, McGraw-Hill, Two Penn Plaza, New York, NY 10121-2298. Or contact your local bookstore.

Mike Meyers' RFID+™ Radio Frequency Identification Certification Passport

1234567890 DOC DOC 01987

ISBN-13: Book P/N 978-0-07-226367-1 and CD P/N 978-0-07-226368-8
of set 978-0-07-226366-4

ISBN-10: Book P/N 0-07-226367-9 and CD P/N 0-07-226368-7
of set 0-07-226366-0

Sponsoring Editor Tim Green	**Indexer** Valerie Robbins
Project Editor Patty Mon	**Production Supervisor** Jim Kussow
Acquisitions Coordinator Jennifer Housh	**Composition** Apollo Publishing Services
Technical Editor Joe Yacura	**Illustration** Lyssa Wald
Copy Editor Lisa Theobald	**Art Director, Cover** Jeff Weeks
Proofreader Stefany Otis	

CompTIA Authorized Quality Curriculum

The logo of the CompTIA Authorized Quality Curriculum (CAQC) program and the status of this or other training material as "Authorized" under the CompTIA Authorized Quality Curriculum program signifies that, in CompTIA's opinion, such training material covers the content of CompTIA's related certification exam.

The contents of this training material were created for the CompTIA RFID+ exam covering CompTIA certification objectives that were current as of April 2007.

CompTIA has not reviewed or approved the accuracy of the contents of this training material and specifically disclaims any warranties of merchantability or fitness for a particular purpose. CompTIA makes no guarantee concerning the success of persons using any such "Authorized" or other training material in order to prepare for any CompTIA certification exam.

How to Become CompTIA Certified

This training material can help you prepare for and pass a related CompTIA certification exam or exams. In order to achieve CompTIA certification, you must register for and pass a CompTIA certification exam or exams.

To become CompTIA certified, you must do the following:

1. Select a certification exam provider. For more information please visit http://www.comptia.org/certification/general_information/ exam_locations.aspx

2. Register for and schedule a time to take the CompTIA certification exam(s) at a convenient location.

3. Read and sign the Candidate Agreement, which will be presented at the time of the exam(s). The text of the Candidate Agreement can be found at http://www.comptia.org/certification/general_information/ candidate_agreement.aspx

4. Take and pass the CompTIA certification exam(s).

For more information about CompTIA's certifications, such as its industry acceptance, benefits, or program news, please visit www.comptia.org/certification.

CompTIA is a not-for-profit information technology (IT) trade association. CompTIA's certifications are designed by subject matter experts from across the IT industry. Each CompTIA certification is vendor-neutral, covers multiple technologies, and requires demonstration of skills and knowledge widely sought after by the IT industry.

To contact CompTIA with any questions or comments, please call (630) 678-8300 or e-mail questions@comptia.org.

Thanks to everyone who supported me in this effort. Special thanks to those of you who stood by me to ensure that the "right thing" happened—you know who you are. This book represents a second chance for me, to believe that there are honest people in the world, and that a man's word still does mean something. In the end, truth will remain, all else will fall away. The team of folks that helped pull this together worked very hard and I am certain they will all go far. There is no shame in not knowing; the shame lies in not finding out.

—Mark Brown

I dedicate this book to my wife, Vriti, who encouraged me to write.

—Sam Patadia

I would like to dedicate this book to my wife, Anu, and kids, Archit, Arpit & Ankit, for their love and support over the years. And to my parents, thank you. None of this would have been possible without your love, guidance, and care.

—Sanjiv Dua

Contents

Acknowledgments

Thanks go out to Sanjiv Dua for constantly pushing me to do more, Sam Patadia for helping with the "heavy lifting," and Eva Zeisel for all the hard work and dedication she showed throughout this enormous effort.

—Mark Brown

I would like to thank RFID4U team and business partners for their support and input. I would also like to thank Eva Zeisel, Senior Director of Knowledge Solutions, RFID4U, and Jim Briggs for editing and proofreading this book.

—Sam Patadia

I would like to thank RFID4U team for their support over the years. I would also like to thank Eva Zeisel, Senior Director of Knowledge Solutions, RFID4U, for her continuous support, especially in the final stages of this work, where she helped edit and proofread almost the entire book. And, finally, to all my friends and RFID4U business partners: Thank you for all the incredible support!

—Sanjiv Dua

Check-In

May I See Your Passport?

What do you mean you don't have a passport? Why, it's sitting right in your hands, even as you read! This book is your passport to a very special place. You're about to begin a journey, my friend, a journey toward that magical place called *certification*! You don't need a ticket, you don't need a suitcase—just snuggle up and read this passport—it's all you need to get there. Are you ready? Let's go!

Your Travel Agent: Mike Meyers

Hello! I'm Mike Meyers, president of Total Seminars and author of a number of popular certification books. On any given day, you'll find me replacing a hard drive, setting up a Web site, or writing code. I love every aspect of this book you hold in your hands. It's part of a powerful new book series called the Mike Meyer's Certification Passports. Every book in this series combines easy readability with a condensed format—in other words, it's the kind of book I always wanted when I went for my certifications. Putting a huge amount of information in an accessible format is an enormous challenge, but I think we have achieved our goal and I am confident you'll agree.

I designed this series to do one thing and only one thing—to get you the information you need to achieve your certification. You won't find any fluff in here. The authors pack every page with nothing but the real nitty gritty of the RFID+ Certification exam. Every page has 100 percent pure concentrate of certification knowledge! But we didn't forget to make the book readable, so I hope you enjoy the casual, friendly style.

My personal e-mail address is mikem@totalsem.com. Please feel free to contact me directly if you have any questions, complaints, or compliments.

Your Destination: RFID+ Certification

This book is your passport to CompTIA's RFID+ Certification, the vendor-neutral industry-standard certification for Radio Frequency Identification technicians, the folks who build and fix RFID systems. To get RFID+ certified, you need to pass the RFID+ exam from CompTIA. The RFID+ exam concentrates on a technician's ability to install, maintain, repair, and troubleshoot RFID-related hardware and software.

RFID+ certification can be your ticket to the next step in your IT career or simply an excellent step in your certification pathway. This book is your passport to success on the RFID+ Certification exam.

Your Guides: Mike Meyers, Mark Brown, Sam Patadia, and Sanjiv Dua

You get a very talented group of tour guides for this book in Mark, Sam, and Sanjiv. I've written numerous computer certification books—including the best-selling *All-in-One A+ Certification Exam Guide*—and written significant parts of others, such as the first edition of the *All-in-One Network+ Certification Exam Guide*. More to the point, I've been working on PC and computer technology and have been teaching others how to make and fix PCs for a *very* long time, and I love it! When I'm not lecturing or writing about PCs, I'm working on PCs or spanking my friend Scott in Half Life or Team Fortress—on the PC, naturally! These guys are all experts in RFID and I can't think of a better team to help you prepare for your exam.

Mark Brown is VP of professional services at RFID4U, a global provider of RFID education and advisory services with RFID design, construction, and integration projects throughout the United States, Europe, and Asia. Mark is leading cutting-edge RFID deployments. He and the team of experienced consultants he leads are industry recognized and trusted subject matter experts known for their participation in major industry initiatives, such as Auto-ID Labs and EPCglobal workgroups. They have authored well-publicized white papers and three best-selling RFID certification books and speak at major trade shows and industry events. RFID4U partners with the best RFID manufacturers, service providers, and laboratories throughout the world, demonstrating cutting-edge technology to solve challenges throughout diverse organizations in all industry verticals.

Sam Patadia is VP of engineering at RFID4U. As a subject matter expert in RFID, he provides consulting services for the companies implementing RFID systems. He has developed several basic- to advanced-level training courses for RFID hardware and software and has taught those courses all over the world. As an RFID evangelist, Sam is a frequent speaker at many universities, major trade shows, and industry events. He has spent more than 25 years as a consultant in design and development of electromechanical systems and software. Some of the companies for which he provided consulting services include General Electric, Westinghouse, Simulation Associates, ROLM (Siemens), Tandem (HP), Informix, U.S. Postal Service, and Alza Corporation. He received a B.S. and M.S. in engineering from North Carolina A & T State University.

Sanjiv Dua is a serial entrepreneur with 17-plus years of industry experience, who has provided the vision, management, and marketing strategy underlying a number of successful information technology product and service companies. Prior to RFID4U (world leader in vendor-neutral RFID learning solutions), Sanjiv was senior regional director at Inteliant Corporation, part of a publicly traded IT solutions company, where he provided overall direction and management for a staff of more than 175. Sanjiv also has held the post of country manager for Allied Telesyn, where he managed the company's business in Southeast Asia. His previous career experience includes senior-level national and international business development positions at Aquas, Sprint, and Tata-Elxsi in United States, Europe, Singapore, and India. He has a B.S. in computer engineering and has attended several management courses.

About the Tech Editor

Joseph A. Yacura is co-founder and chief strategist of Supply Chain Management, LLC, a supply chain and cost management consulting firm. The company provides consulting services related to supply chain management, accounts payable, finance, change management, quality management, and information systems. Joseph has spent more than 30 years assisting companies in utilizing processes, quality, financial controls, and technology to enhance their competitive posture. During his early career he spent 10 years with IBM in various management positions. As a senior vice president and/or chief procurement officer at Pacific Bell, American Express, Bank of America, and InterContinental Hotels Group, he gained insight to several different vertical industries. Joseph received his undergraduate degree from the State University of New York at Oswego. He received an M.B.A. and M.S. degree from Binghamton University and an M.Q.M. degree from Loyola University. He has also completed the Senior Executive Program at Stanford University.

About LearnKey

LearnKey provides self-paced learning content and e-learning solutions to enhance personal skills and business productivity. LearnKey claims the largest library of rich streaming-media training content that engages learners in dynamic media-rich instruction complete with video clips, audio, full-motion graphics, and animated illustrations. LearnKey can be found on the Web at www.LearnKey.com.

Why the Travel Theme?

The steps to gaining a certification parallel closely the steps to planning and taking a trip. All of the elements are the same: preparation, an itinerary, a route, and even mishaps along the way. Let me show you how it all works.

This book is divided into 10 chapters. Each chapter begins with an *Itinerary* that provides objectives covered in each chapter, and an *ETA* to give you an idea of the time involved learning the skills in that chapter. Each chapter is broken down by objectives, either those officially stated by the certifying body or our expert take on the best way to approach the topics. Also, each chapter contains a number of helpful items to bring out points of interest:

Exam Tip
Points out critical topics you're likely to see on the actual exam.

Travel Assistance
Shows you additional sources, such as books and Web sites, to give you more information.

Local Lingo
Describes special terms in detail, in a way you can easily understand.

Travel Advisory
Warns you of common pitfalls, misconceptions, and downright physical peril!

The end of the chapter gives you two handy tools. The *Checkpoint* reviews each objective covered in the chapter with a handy synopsis—a great way to review quickly. Plus, you'll find questions to test your newly acquired skills.

CHECKPOINT

But the fun doesn't stop there! After you've read the book, pull out the CD and take advantage of the free practice questions. Use the full practice exam to hone your skills and keep the book handy to check answers.

When you're acing the practice questions, you're ready to take the exam. *Go get certified!*

The End of the Trail

The IT industry changes and grows constantly, *and so should you*. Finishing one certification is just a step in an ongoing process of gaining more and more certifications to match your constantly changing and growing skills. Read the Career Flight Path at the end of the book to see where this certification fits into your personal certification goals. Remember, in the IT business, if you're not moving forward, you are way behind!

Good luck on your certification! Stay in touch!

Mike Meyers
Series Editor
Mike Meyers' Certification Passport

RFID Basics

	NEWBIE	SOME EXPERIENCE	EXPERT
ETA	1 hour	30 minutes	15 minutes

1

Radio Frequency Identification (RFID) technology belongs to a broader group of technologies known as *Auto Identification* (Auto-ID). This chapter will take a brief look at some of the Auto-ID technologies and will introduce RFID technology, define some basic RFID terminology, review a brief history of RFID, discuss organizations that create RFID standards, and provide a few examples of systems using RFID technology.

Objective 1.1   Auto-ID Technologies

An Auto-ID technology is anything that collects data about the objects and enters that data into a database without human intervention. Auto-ID technologies are everywhere, quietly and efficiently doing thousands of mundane jobs. The one big job where Auto-ID makes a natural fit is in answering some of the big questions of commerce: "What is it?", "Where is it?", and "What about it?"—primarily the identification and tracking of boxes, people, animals, you name it! Compared to humans, Auto-ID technologies identify and track faster, more accurately, and at a reduced overall cost. RFID is only one of many types of Auto-ID technologies. Other Auto-ID technologies include Magnetic Ink Character Recognition (MICR), magnetic strip, voice recognition, biometrics, and barcodes.

MICR reads ink-printed characters, such as those that often appear at the bottom of personal checks. The checks must be properly oriented and presented to the MICR reader one at a time. Magnetic strips are used on credit and debit cards and also require a proper orientation and physical contact between the card and the reader. Barcodes consist of a series of black bars and white spaces of varying widths. Several hundred different types of barcodes are used, with the most common being the Uniform Product Code (UPC), which is used extensively by the retail industry. Barcodes require a line of sight and proper orientation of the barcodes relative to the scanner. Voice recognition is used by order-picking applications in distribution centers (DCs). In order picking, voice recognition has a big advantage over barcode identification. It allows hands-free and eyes-free order picking and does not require alignment of labels to readers. Biometrics, such as fingerprint and retinal scans, are used to identify people. Many of the latest computers use fingerprints to identify the user. In many highly secured locations, entry permits are granted using retinal scans. Retinal scanning has also been used to identify cattle.

So, with all these Auto-ID technologies, why should yet another technology like RFID suddenly become so popular? It all boils down to one thing: radio waves. RFID encompasses technologies that use electromagnetic (radio) waves, part of electromagnetic spectrum, to identify individual items, places, animals, or people. RFID can be appropriately implemented for many different uses. The most common is to use an identifying number (sort of a name) that uniquely identifies an object, place, animal, or person. The number is stored on an integrated circuit (IC) that is attached to an antenna. Together, the IC and the antenna are called an *RFID transponder* or *tag*. The tag is attached to the object, place, animal, or person to be identified. A device called the *interrogator* or *reader* communicates with the tag and is used to read the identifying number from the tag. The reader feeds the number it reads into an information system, which stores the number in its database or searches its database for the number and returns information stored therein about the object, place, animal, or person. The major difference between various Auto-ID technologies is in how the identifying number is stored and retrieved.

Travel Advisory

Almost all RFID interrogators, though they are called readers, also *write* data to the tag.

Among the Auto-ID technologies mentioned so far, the barcode is closest to RFID; therefore, the two technologies are compared here to help you understand their relative advantages and disadvantages. Two major types of barcodes are linear and matrix.

Linear barcodes use black-and-white stripes of different width to encode a number, while matrix barcodes use two-dimensional arrays of black-and-white patches to encode information. A scanner using a laser reads the information encoded in barcodes and provides it to the information system.

RFID is not necessarily "better" than barcode. The two are different technologies that have different, yet sometimes overlapping, applications. The big difference between the two is that barcodes require a line-of-sight, while RFID does not. A barcode scanner has to "see" the barcode to read it, but the RFID reader can read the tag while it is optically hidden. The barcode also needs to be properly oriented toward a scanner for it to be read. RFID tags can be read as long as they are within the range of a reader, regardless of their orientation.

Barcodes have other shortcomings. If a label is ripped or soiled, the item cannot be scanned. Standard barcodes identify only the manufacturer and product, not the unique item. So, for example, the barcode on one juice carton is the same as every other, making it impossible to identify which one might pass its expiration date first. Many people think that RFID will replace barcodes, but if this happens, it will take several years since barcodes is a mature technology and currently has a very low cost of implementation compared to RFID.

Table 1-1 lists advantages and disadvantages of RFID and barcode.

 **Objective 1.2** ## Brief History of RFID

RFID is called a *new technology*, but it is actually older than barcodes. The technology that forms the basis for RFID was first developed during World War II to identify airplanes. At that time it was called *friend-or-foe* technology, and a greatly modernized version is still used in aircraft identification today. Barcodes were invented in the late 1940s but did not see substantial use until the

TABLE 1.1	Advantages and Disadvantages of RFID and Barcode
RFID	**Barcode**
No line of sight required	Line of sight required
Uniquely identifies items, cases, pallets	Identifies only item category
Item orientation to reader not important	Requires proper orientation
Simultaneous identification	Scans only single item at a time
Dynamic read/write capability	No write capability, static information
Can be used in harsher environment	Soiled labels are difficult to read
More data storage capacity	Limited data storage capacity
Worldwide standards still in process	Worldwide standards in place
More expensive: $0.10-plus cost to attach	Cheaper to produce: $0.001
Currently requires two steps: tag creation and tag attachment	Single step: can be easily printed on boxes during manufacturing

late 1960s and early 1970s. However, barcode's low price compared to RFID made it a much more attractive option for Auto-ID usage at that time. Well, at least if you were not trying to identify an airplane! But, as the cost of RFID slowly dropped, industry began to use it for more applications. Since 1979, RFID has been in use to identify and track animals. In 1994, all rail cars in the United States used RFID tags for identification. A recent surge has occurred in RFID technology research, manufacturing, and usage due to the advances in semiconductor manufacturing, which has reduced the cost of RFID, making their use economically feasible in supply-chain and other applications where tags are attached to objects not normally returned to the manufacturer.

RFID systems in the past used proprietary technologies; no worldwide open standards existed. There was little or no interconnectivity between different RFID vendor's products. Every vendor had their own readers, tags, signals, and equipment—and nothing worked together. This lack of interoperability made it challenging for companies to adopt RFID and constrained the deployment of RFID technology in the global supply chain. However, that problem is quickly disappearing due to new standards. Since 2006, many international and industry organizations have created open standards. RFID equipment manufacturers have started selling tags and readers that conform to these new standards, resulting in an increase in the supply and a reduction in the price of RFID devices, which in turn has accelerated RFID technology adoption.

Table 1-2 shows a timeline of some of the important events in RFID.

TABLE 1.2	Important Events in RFID History
Year	**Event**
1948	Harry Stockman—Communications by means of reflected power
1950s	Harris Patent—Radio transmission systems with Modulatable Passive Responder
1973	Cardullo patent for passive RFID
1975	Los Alamos Scientific Laboratory—RFID research released to public
1979	Animal tagging
1987	Motor vehicle toll collection in Norway
1991	Association of American Railroads standards
1994	All US railcars RFID enabled
1999	Massachusetts Institute of Technology Auto-ID center founded
2003	EPCglobal system Version 1.0
2005	US Department of Defense and Wal-Mart mandates

Objective 1.3   **RFID Systems**

R FID may only consist of a tag and a reader but an RFID system comprises many other technologies, such as computer, network, Internet, wireless devices, and software, all working with the RFID devices to create a complete solution. A typical RFID system is divided into two layers: the physical layer and Information Technology (IT) layer.

The physical layer consists of the following:

- One or more RF tags
- One or more interrogators (readers)
- One or more reader antennas
- Deployment environment

The IT layer consists of the following:

- One or more host computers connected to readers (directly or through a network)
- Appropriate software (device drivers, filters, middleware, databases, and user applications)

Figure 1-1 provides a bird's-eye view of the RFID system, showing tags, readers, network, computers with software applications, and people all interacting to monitor and control business processes. The bidirectional mode of

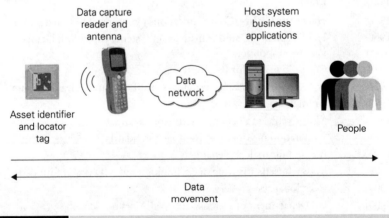

FIGURE 1.1   Bird's-eye view of an RFID system

data movement among various parts of the RFID system is depicted at the bottom of the figure. Data may be read from or written to the tag during a business process. For example, a number may be read from the tag attached to a case of goods passing through the shipping dock, while data may be written to the tag attached to a part moving from one workstation to the other during the manufacturing process.

Figure 1-2 shows the physical layer of an RFID system: a tag, an antenna, a reader, and the deployment environment. The deployment environment consists of an interrogation zone (IZ)—the space in which a reader antenna emits radio waves through which the tags pass—and objects in the vicinity of the IZ. The deployment environment is included in the physical layer because the performance of the RFID reader and tag is greatly affected by various characteristics of the deployment environment. The radio frequency interference within the deployment space and type, size, and shape of objects located within the deployment space affect the read performance of tags.

All RFID systems require IT layer components. The IT layer consists of various computer systems, networks, databases, and application software. RFID software is divided into two groups: *middleware* and *enterprise applications*. Middleware directly interacts with the RFID physical layer, collects data from readers, attaches business process information to data, stores data, and supplies data to enterprise applications in their native formats. It also manages, monitors, and configures hardware. Middleware forms a conduit between the enterprise applications used to manage business processes and hardware components. Enterprise applications, also called business applications, use middleware to gather data from RFID readers. This data is then used to manage business. For example, data received from RFID readers at shipping dock may be used to create invoices and bill customers.

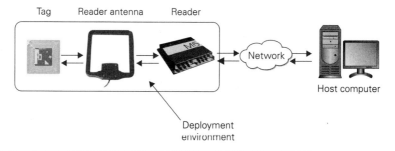

FIGURE 1.2   The physical layer of the RFID system

Exam Tip

The CompTIA RFID+ Certification Exam includes only the physical layer of the RFID technology. Therefore, only the physical components of the RFID system are discussed in this book.

Many varieties of RFID tags, readers, and antennas are available on the market. The RFID system designer selects them according to the requirements of the objects to be tagged, the distance at which the tags are to be read, the business processes during which tags are read, the speed of the tagged objects through the IZ, and the number of tags in the IZ. RFID tags come in many different shapes and sizes, operate at different frequencies, use different protocols, obtain power from different sources, can be written to once or several times, and currently cost anywhere from ten cents to several dollars. RFID readers are designed to operate at different frequencies, protocols, and power levels. RFID antennas also come in different sizes, shapes, frequencies, and with different radiation patterns. All of these aspects are discussed in detail in the chapters that follow.

Objective 1.4   **RFID Standards and Regulations**

To be widely accepted, any technology requires some sort of standards and regulations that provide guidelines for designing, manufacturing, and using the technology. Standards are created by industry consortiums such as EPCglobal or standards-creating organizations such as the International Organization for Standardization (ISO). Standards help provide interoperability among vendors' products, increased demand for products, and reduced costs. Regulations are created by governmental agencies and help improve safety of the technology and create an orderly development and deployment environment in which several similar technologies can operate simultaneously.

Two standards developing organizations most instrumental in getting different RFID solutions to work together are EPCglobal and ISO. EPCglobal is a not-for-profit industry organization. The acronym *EPC* stands for *Electronic Product Code*. Some of the standards developed by EPCglobal include Gen 2 (or Generation 2) and ALE (Application Level Events). The Gen 2 standard defines rules for communication between the tag and the reader, while the ALE standard defines rules for data collection and filtering and reader management and monitoring. ISO has approved the Gen 2 standard as ISO 18000-6C. EPCglobal is currently working on development of the EPCglobal network, which will provide a standardized way to collect, store, and distribute data among various organizations in the supply chain.

Objective 1.5 Benefits of RFID

One of the benefits of the RFID system is reduced human intervention. RFID helps collect data automatically while business processes are being performed. No special actions are necessary to collect data. This automation improves data quality, reduces data collection time, captures data in real-time, and reduces costs associated with poor data quality. Time and cost savings also occur because boxes need not be opened to scan items to collect data. Inventory information can be collected in real-time and items can be tracked in real-time. RFID can also be utilized to verify the origin of an item and thus help detect counterfeit items. With the reduced cost of collecting data, businesses can afford to collect data at more locations throughout a process. This provides finer visibility in the business process, helps fine-tune the business processes, and helps identify inventory losses due to theft and damage. You can write data to a tag so the information travels with the item and can be used by downstream business processes. You can also uniquely identify each instance of an item and scan multiple items at one time at a high speed, even in harsh environments.

Objective 1.6 RFID Applications

RFID technology can be used in a wide variety of industries that need to track objects, persons, or animals. A few of the examples of applications are listed here:

* **Manufacturing**   Track work in progress to identify and eliminate bottlenecks, and track finished goods, inventory levels, and locations

- **Automotive**   Engine immobilization system
- **Supply chain**   Shipping and receiving, warehousing, retail outlet, inventory management
- **Pharmaceutical**   Pedigree management, product authentication, documents for FDA
- **Healthcare**   Track patients, equipment, and services
- **Library and video store**   Track assets and rentals
- **Access control**   Track security badges
- **Finance**   Cashless payments
- **Entertainment**   Hospitality, amusement park, event management, and ticket sales and verification
- **Inventory**   Recall and return management
- **Document management**   Track documents in lawyers' offices, hospitals, government offices
- **Transportation management**   Rail car and truck tracking, toll collection, vehicle theft detection, vehicle speed tracking
- **Sports**   Track timing
- **Livestock and wildlife**   Track animals
- **Airline industry**   Track baggage

Following are a few examples of RFID systems.

Building Access System   The first example is a building access system that uses badges with an imbedded RFID tag. Many organizations use RFID-enabled access badges. The badge contains an RFID tag on which an identifying number is stored. When the badge holder wants to enter the protected area, he holds the badge near a special reader near the door. The reader reads the number from the badge and sends it to the computer system. The computer system compares that number with the stored information and determines whether the person holding the badge has access privileges to the area. If the badge holder is allowed access to the area, the door is unlocked.

Supply Chain System   An RFID system can also be used in a supply chain. Consider a simple supply chain consisting of three companies: a manufacturer, a distribution center (DC), and a retailer. The manufacturer produces some items with attached RFID tags encoded with unique numbers. When the manufacturer ships items to the DC, an RFID reader at its shipping dock automatically reads the numbers from the tags on the items. These numbers, along with other information such as read time and dock location, are stored in the company's database system. While the items are being shipped, the manufacturer sends an advance ship notice (ASN) to the DC. The ASN contains, along with other data, a list of numbers read from all the tags of the items shipped.

When the items arrive at the receiving dock of the DC, an RFID reader reads the numbers from the tags attached to the items. These numbers are compared to those on the ASN from the manufacturer. If the numbers do not match, an exception is generated and the discrepancy is automatically reported to the manufacturer and to appropriate people within the DC. If numbers match, the shipment is accepted. The numbers of the items received are recorded in the DC's database. When the DC ships the items to the retailer, the numbers from the attached tags on the items being shipped are read at the shipping dock door. The DC's inventory system is updated, items shipped to the retailer are recorded, and an ASN is sent to the retailer.

When the retailer receives the items, a process similar to the one performed at the receiving dock of the DC is performed at the retailer's receiving dock. Now all three companies have complete records of what, when, and to whom items were shipped and what, when, and from whom items were received. The manufacturer, if authorized by the DC and the retailer, can look in their database inventory of the items it manufactured and can use that information to plan its future manufacturing schedule and order raw materials. All three companies can have access to all the databases and use that information for optimizing the inventory to carry, the number of items to order, and the number of items to manufacture. This creates an efficient supply chain and lowers the cost for all parties involved. This data can also be used to track product quality, support warranties, product recalls, and other information.

CHECKPOINT

✔ **Objective 1.1: Auto-ID Technologies** Auto-ID technologies such as RFID, barcodes, MICR, magnetic strip, and biometrics are used to identify items of interest faster, more accurately, and at reduced cost. RFID uses IC to store data and radio waves to transmit them, while barcodes use black-and-white stripes or patches to store data and light waves to transmit them. Both of these technologies are very similar in use once the data is read from the data storage medium. The differences between them are in how the data is stored and retrieved, the amount of data that can be stored, and whether the data can be rewritten or not.

✔**Objective 1.2: Brief History of RFID**   RFID technology has existed since the 1950s and has been in use since the late 1970s. Recent advances in semiconductor manufacturing have reduced the cost of RFID to a point that it is now economical to use RFID on a large scale.

✔**Objective 1.3: RFID Systems**   RFID systems consist of physical and IT layers. The physical layer includes RF tags, interrogators, antennas, and deployment environment, while the IT layer includes computer hardware and software. RFID tags and readers are designed to operate at various frequencies and protocols. The performance of the RFID system is affected by types of tags, readers, and antennas used and the RF characteristics of the deployment environment.

✔**Objective 1.4: RFID Standards and Regulations**   EPCglobal and ISO have recently developed standards for the design and operation of RFID devices, such as tags and readers. The standardization has helped to accelerate the implementation of RFID systems. All RFID systems must comply with the government regulation in the country of use.

✔**Objective 1.5: Benefits of RFID**   RFID systems reduce human intervention in the business processes and thus help improve efficiency and reduce cost. They also provide data in real-time and help detect counterfeit items and shrinkage.

✔**Objective 1.6: RFID Applications**   RFID systems can be used in any application where an item, an animal, or a person needs to be quickly and efficiently identified. RFID systems are used in a variety of areas, such as manufacturing, supply chain, healthcare, library, asset tracking, and automotive.

REVIEW QUESTIONS

1. Which of the following is used by RFID technology to communicate between tag and reader?

   A. Light

   B. Gamma ray

   C. Electric current

   D. Electromagnetic wave

2. Which of following is an advantage of barcode over RFID?

   A. Can read without line-of-sight

   B. Cheaper to implement

   C. Can read multiple items at a time

   D. Can be used in a harsher environment

3. The EPCglobal Gen 2 standard has been approved as what?

   A. ISO 18000-6A

   B. ANSI 3.2

   C. ETSI 200-302

   D. ISO 18000-6C

4. Which of the following is not a part of the RFID physical layer?

   A. Communications network

   B. Deployment environment

   C. Tag

   D. Antenna

5. RFID technology is not appropriate for tracking what?

   A. People

   B. Satellites

   C. Objects in a supply chain

   D. Cattle

6. RFID is not economical for tracking what?

   A. Cases of cereal boxes

   B. Pallets of shampoo bottles

   C. Patients in hospital

   D. Individual cans of soda

REVIEW ANSWERS

1. **D**   Electromagnetic waves are used by RFID technology to communicate between tag and reader.

2. **B**   Barcodes are cheaper to implement. RFID tags cost at least 100 times more than barcodes.

3. **D**   ISO 18000-6C was approved by the ISO for the EPCglobal Gen 2 standard.

4. **A**   A communications network is not a part of the RFID physical layer.

5. **B**   Satellites are too far away to be tracked using RFID tags.

6. **D**   Individual cans of soda currently cost less than most RFID tags; therefore, it is not economical to use RFID tags to track them.

The RF in RFID

	NEWBIE	SOME EXPERIENCE	EXPERT
ETA	6 hours	3 hours	1 hour

RFID systems use radio waves to exchange information between RFID transponders, or tags, and interrogators or readers. How radio waves behave under various conditions in the RFID interrogation zone (IZ) affects the performance of the RFID system. Radio waves propagate from their source and reach the receiver. During their travel, they pass through different materials, encounter interference from their own reflection and from other signals, and may be absorbed or blocked by various objects in their path. The material of the object to which the tag is attached may change the property of the tag.

To understand how all these phenomena affect the performance of the RFID system, you need to understand how radio waves are used and propagated. The first part of this chapter discusses how radio waves travel and how their effectiveness changes during this travel. The second part discusses antennas, the devices that transmit and receive the radio signals. You will study different characteristics of an antenna such as its gain, directivity, polarization, and impedance. You will also look at various types of antennas and their usage and the types of antenna cable and connectors that are used to connect antennas to RFID interrogators.

 **Objective 2.1** **Radio Frequency Basics**

Radio frequencies constitute a small portion of a larger electromagnetic spectrum. The total electromagnetic spectrum includes other higher frequency waves such as light, ultraviolet ray, X-ray, and gamma-ray, which are not discussed in this book. You will study the various ways energy in the spectrum propagates and you will learn the various terms used in radio frequency system design.

Electromagnetic Spectrum

In an RFID system, radio waves in a particular frequency range are used to encode and transfer information between an RFID interrogator(s) and each RFID tag. The entire electromagnetic spectrum includes gamma-rays, X-rays, ultraviolet light, visible light, infrared light, microwaves, and radio waves. The difference between these various parts of the electromagnetic spectrum is their *wavelength*, or *frequency*. The energy radiated by radio waves is called *electromagnetic radiation*.

A wave is a distortion in a material or medium. While the wave passes through the material or medium, the individual parts of the material cycle back and forth or up and down. All the waves have similar characteristics and follow the same laws and principles. The characteristics of a waveform are *wavelength*, *amplitude*, *velocity*, and *frequency*. All periodic waveforms have these common characteristics. Figure 2-1 shows a wave with two complete cycles, its amplitude, its wavelength, and points where peaks and valleys occur.

Local Lingo

Sinusoidal wave   A continuous, uniform wave with a constant frequency and amplitude whose waveform is the same as that of a trigonometric sine function.

The *amplitude* of a wave is defined as half the distance between the values at positive and negative peaks. Amplitude relates to loudness in sound and brightness in light.

The *wavelength* of a wave is the distance between a point on one cycle and the identical point on the next cycle, as shown in Figure 2-1, in which points have been selected at the crest of the wave. The typical units of measure of a wavelength are meters or feet. Wavelength is represented by the Greek letter λ (*lambda*).

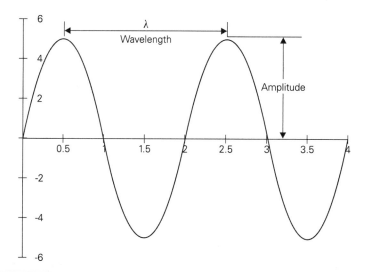

FIGURE 2.1   A sinusoidal wave

The *frequency* of waves is the number of times the identical points on the wave, crests, or peaks, pass any fixed point during 1 second. Frequency is measured in cycles per second. Cycles per second is called *Hertz* (Hz) in honor of German scientist Heinrich Rudolf Hertz. One cycle per second is equal to 1 Hz. *Frequency* is represented by the letter *f*.

The *velocity* of the wave is the measure of how fast any point on the wave is moving past a fixed point. Typical units of velocity are meters/second or miles/second. Velocity is represented by a lowercase letter *v*.

The relationship between velocity, frequency, and wavelength is shown by the following equation:

Velocity (*v*) = wavelength (λ) × frequency (*f*) or
$$v = \lambda \times f$$

Using this equation, you can calculate frequency from wavelength or vice versa.

Now let's go back to the electromagnetic spectrum. Figure 2-2 shows various parts of the electromagnetic spectrum and their wavelengths. Gamma-rays are at the top, and radio waves are at the bottom. The radio waves are further divided into groups called *low frequency* (LF), *high frequency* (HF), *ultra high frequency* (UHF), and *microwave*. RFID technology uses radio waves within all of these ranges.

Electromagnetic radiation can be described in terms of a stream of photons. Photons are massless particles, each traveling in a wavelike pattern and moving at the speed of light. Each photon contains a certain amount of energy. The only difference between the various types of electromagnetic radiation is the amount of energy found in the photons. Radio waves have photons with the lowest energy, microwaves have a little more energy than radio waves, infrared has still more energy, and then visible, and then ultraviolet, X-rays, and gamma-rays. The electromagnetic spectrum can be expressed in terms of energy, frequency, or wavelength, all of which are mathematically related. The main reason for using one of these three different ways to describe waves is for convenience—to avoid handling large numbers. Since radio waves have photons with very low energy, scientists and engineers use only frequency and wavelength to describe them.

Electromagnetic Propagation

An electromagnetic wave moves or propagates in a direction that is at right angles to the vibrations of both the electric and magnetic oscillating field vectors, carrying energy from its radiation source to an undetermined final

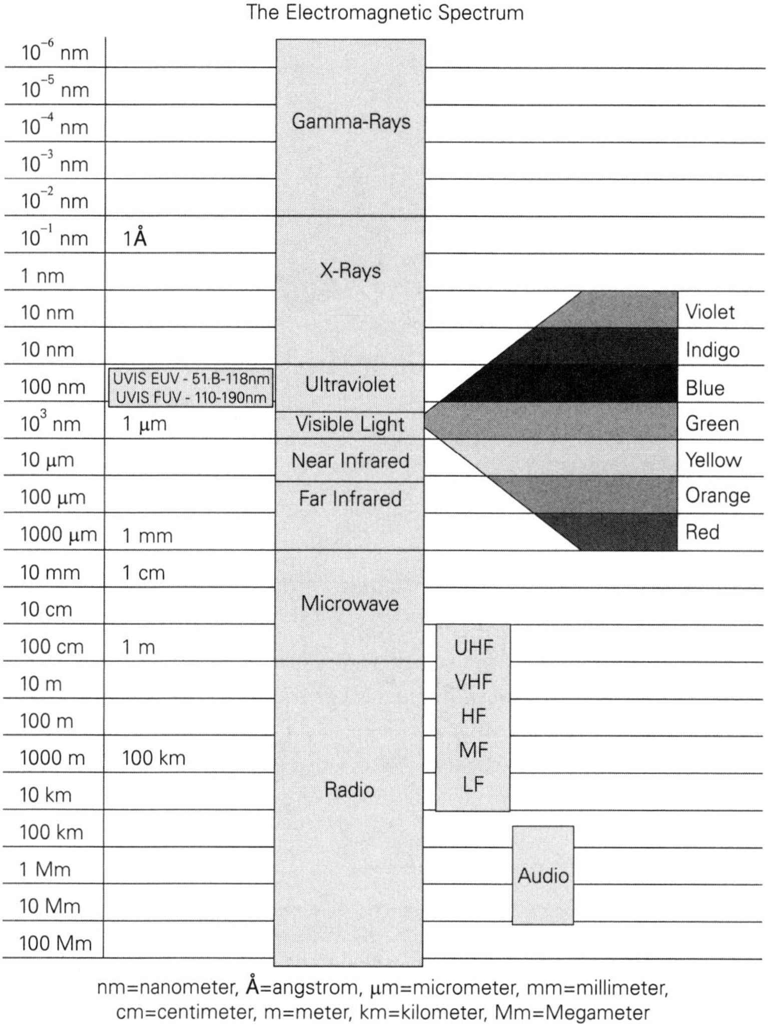

The Electromagnetic Spectrum

10^{-6} nm			
10^{-5} nm			
10^{-4} nm		Gamma-Rays	
10^{-3} nm			
10^{-2} nm			
10^{-1} nm	1Å		
1 nm		X-Rays	
10 nm			Violet
10 nm			Indigo
100 nm	UVIS EUV - 51.B-118nm / UVIS FUV - 110-190nm	Ultraviolet	Blue
10^3 nm	1 µm	Visible Light	Green
10 µm		Near Infrared	Yellow
100 µm		Far Infrared	Orange
1000 µm	1 mm		Red
10 mm	1 cm		
10 cm		Microwave	
100 cm	1 m		UHF
10 m			VHF
100 m			HF
1000 m	100 km		MF
10 km		Radio	LF
100 km			
1 Mm			Audio
10 Mm			
100 Mm			

nm=nanometer, Å=angstrom, µm=micrometer, mm=millimeter,
cm=centimeter, m=meter, km=kilometer, Mm=Megameter

FIGURE 2.2 The electromagnetic spectrum

destination. As shown in Figure 2-3, the electric and magnetic fields are mutually perpendicular.

All electromagnetic waves travel at the same speed in vacuum—that is, at the speed of light, 299,792,458 meters per second (m/sec). The speed of light is designated by the lowercase letter c and is usually approximated to 300,000,000 m/sec. For all our calculations, we will use the value of 300,000,000, or 3×10^8 m/sec.

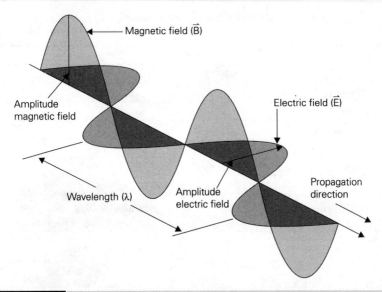

| **FIGURE 2.3** | Electromagnetic wave propagation |

Using the speed of electromagnetic waves as c, we can rewrite the relationship, described earlier, between frequency (f) and wavelength (λ) as follows:

$$c = \lambda \times f \qquad \lambda = \frac{c}{f} = \frac{3 \times 10^8}{f} \qquad f = \frac{c}{\lambda} = \frac{3 \times 10^8}{\lambda}$$

Exam Tip

Know that electromagnetic waves travel at the speed of light, 3×10^8 m/sec, in a vacuum. The speed is almost the same in air.

As electromagnetic waves travel in space, they spread their power in a sphere of increasing radius. This means the same amount of power is spread over the surface of a larger sphere. This reduces the power density at any point as the wave travels away from the source. This phenomenon is called *path loss*. Figure 2-4 shows a transmit antenna TX at the center of a sphere and a receive antenna

RX at the surface of the sphere. The distance between the two antennas is *r* and the effective area of the receive antenna is *A*.

$$Path\ loss = \frac{RX\ effective\ area}{total\ area} = \frac{A}{4 \cdot \pi \cdot r^2}$$

From this equation, you can conclude that the power density (PD) of a radio wave at a distance (r) from its source is inversely proportional to the square of the distance—PD á 1/r². In other words, power attenuates with the square of the distance.

Exam Tip
Know that wavelength = $3{\times}10^8$ m/sec / frequency.

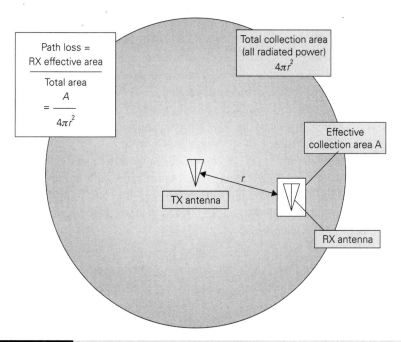

FIGURE 2.4   Electromagnetic wave path loss in free space

Radio Waves

RFID technology uses the radio wave portion of the electromagnetic spectrum; therefore, our discussion of electromagnetic spectrum is limited to the radio wave portion of the spectrum, specifically the radio waves in the range of 100 kHz to 5.8 GHz.

Radio signals are affected in many ways by objects in their path and by the media through which they travel. This means that radio signal propagation is of vital importance to anyone designing or operating a radio system. The properties of the path through which the radio signals propagate govern the level and quality of the received signal. Reflection, refraction, and diffraction of the original signal can occur and may add constructively or destructively to the original signal. The resultant signal is a combination of several signals that have traveled by different paths due to reflection, refraction, and diffraction. In addition, the signals traveling via different paths can be delayed and can distort the resultant signal.

Radio signals propagate in many different ways, including free space propagation, ground wave propagation, ionospheric propagation, and tropospheric propagation. These relate to the effects of the media through which the signals propagate. RFID applications occur at very small distances; therefore, free space propagation is the most important type. In free space propagation, the major factor affecting the signal strength is the distance between the transmitter and the receiver. RFID systems need to have their radio propagation models generated for factory, warehouse, office, or urban situations. Under these circumstances, the free space propagation is modified by multiple reflections, refractions, and diffractions. Despite these complications, it is still possible to generate rough guidelines and models for these radio propagation scenarios.

Within the radio spectrum is an enormous range of frequencies. To categorize and manage the different areas of the spectrum, the radio spectrum is split into many different segments, but RFID technology uses only four of these segments, as shown in the following table. Figure 2-5 shows these frequencies in relation to the entire radio spectrum.

Range	LF	HF	UHF	Microwave
Frequency available	30–300 kHz	3–30 MHz	300–1000 MHz	1–6 GHz
Used for RFID	125–134 kHz	13.56 MHz	433 & 860–960 MHz	2.4 & 5.8 GHz

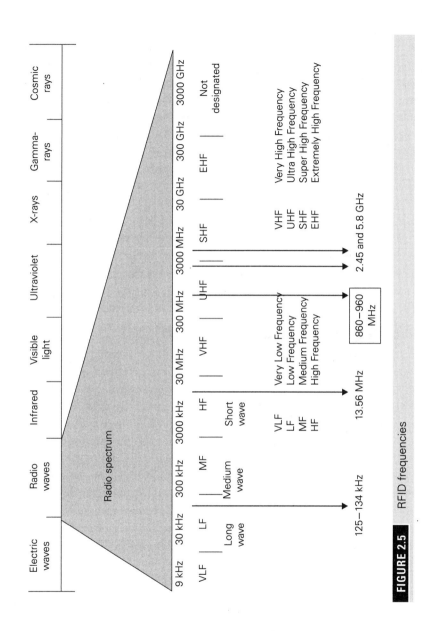

FIGURE 2.5 RFID frequencies

The radio spectrum is an aspect of the physical world, like land, water, and air. Use of radio frequency bands of the electromagnetic spectrum is regulated by governments in most countries, in a process known as *frequency and/or spectrum allocation*. Like weather and internationally traded goods, radio propagation and RF technology do not stop at national boundaries. For economic and technical reasons, governments all over the world have sought to harmonize spectrum allocation standards by regulating various aspects of radio spectrum use.

The following forums and standards bodies have worked and/or contributed to standards for frequency allocation:

- **ITU** International Telecommunication Union
- **CEPT** European Conference of Postal and Telecommunications Administrations
- **ETSI** European Telecommunications Standards Institute
- **CISPR** International Special Committee on Radio Interference

Interference and Multipath

Interference is the superposition of two or more waves resulting in a new wave pattern. Multiple waves may be generated due to various effects of propagation of the original wave and/or completely different waves from other processes. When waves generated due to propagation effects take a different path than the original wave, it is called *multipath*. The phenomenon of wave reflection, diffraction, and scattering all give rise to additional radio propagation paths beyond the direct line-of-sight (LOS) path between the radio transmitter and receiver. Due to multipath, caused by whatever phenomenon, the actual received signal level is a vector sum of all signals incident from any direction or angle of arrival. Some signals add to the direct path, while other signals subtract from the direct signal path. This creates bright and dark spots within the antenna radiation pattern.

Local Lingo

Hot spot *Hot spot* or bright spot refers to spaces within the antenna beam where power density is higher than normal.
Cold spot *Cold spot* refers to spaces within the antenna beam where power density is lower than normal. Hot and cold spots make the antenna radiation pattern look like Swiss cheese.

As radio waves travel, they interact with objects and the media they encounter. This interaction causes reflection, refraction, diffraction, scattering, and absorption of the wave. This interaction also can change the waves' direction of travel or it can block the waves. The change in direction causes the waves to reach areas that would not be possible if the radio signals traveled in a direct line. This has advantages, because the signals can reach an area that is not in direct LOS, and it also has disadvantages, because the reflections of the signal or other signals in the vicinity can interfere with the original signal.

Exam Tip	
Know that multipath causes variations in power density within an antenna beam.	

Reflection

Radio waves, like light waves, are reflected by many surfaces. *Reflection* occurs when a propagating electromagnetic wave impinges upon an object that has very large dimensions compared to the wavelength of the propagating wave. Surfaces of conductive materials reflect signals more than nonconductive materials. Radio signal reflection follows the same law as light wave reflection: the angle of incidence is equal to the angle of reflection. When a radio signal is reflected, some loss of the signal is normally incurred, either through absorption or as a result of some of the signal passing into the medium. In a manufacturing facility or warehouse, metallic objects, such as heavy machinery, shelving, rolling metallic shutters on dock doors, or metallic conveyors, reflect radio waves.

When a radio wave hits a material, some of the power is reflected at the surface and some of the power is transmitted into and possibly through the material. If the material is a conductor such as metal, almost all of the radio power is reflected within the first few atoms of the material. Buildings made of metal, metal-coated glasses, or steel-reinforced concrete, reflect most of the radio energy. When radio waves impinge upon a material that is a dielectric (an insulator), some of the power is reflected at the surface but most of the power travels through the material. How much power travels through a dielectric depends on both the thickness of the material and its attenuation coefficient. Examples of dielectrics are paper, plastic, Teflon, glass, ceramic, and dry wood. Pure water is a good dielectric substance.

Figure 2-6 shows radio wave reflection and multipath created due to reflection.

> **Exam Tip**
>
> Metals reflect radio waves while aqueous liquids absorb radio waves.

Diffraction

Radio signals may also undergo *diffraction*, which occurs when signals encounter an obstacle and tend to travel around them. Diffraction occurs when the radio path between the transmitter and receiver is obstructed by a surface that has sharp irregularities (edges). The secondary waves resulting from the obstructing surface are present throughout the space, even behind the obstacle, giving rise to a bending of waves around the obstacle, even when a LOS path does not exist between transmitter and receiver. At high frequencies, diffraction, like reflection, depends on the geometry of the object, as well as the amplitude, phase, and polarization of the incident wave at the point of diffraction.

Scattering

Scattering occurs when the medium through which the wave travels consists of objects with dimensions that are small compared to the wavelength, and where the number of such objects per unit volume is large. Scattered waves are produced by rough surfaces, small objects, or by other irregularities in the channel. When a radio wave impinges on a rough surface, the reflected energy is diffused

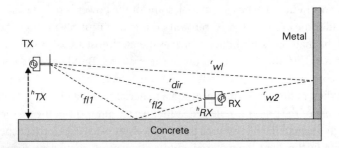

FIGURE 2.6   Radio wave reflection

in all directions due to scattering. Figure 2-7 shows radio wave diffraction and scattering.

Refraction

Refraction is a change in direction and speed of travel when radio waves move from a medium of one refractive index to another with a different refractive index. This is similar to the effect you see when you place a pencil at an angle in a glass of water. The pencil seems to bend. This phenomenon does not have much importance in RFID application.

Fading

Another term frequently used in the RF arena is *fading*. Fading is the variation of signal strength with time. It happens due to time dependent random changes in multipath. A fade is a constantly changing, three-dimensional phenomenon. Fade zones tend to be small, multiple areas of space within an IZ that cause periodic attenuation of a received signal. In other words, the received signal strength will fluctuate downward, causing a momentary, but periodic, degradation in radio signal quality. One may visualize signal strength with the IZ as a Swiss cheese, where holes in the cheese represent fade zones. The fading effect grows worse at greater distances from the interrogator antenna. It is impossible to estimate signal strength accurately at various points in an IZ due to the random nature of fades.

dB and dBm

Signal levels in radio technology vary by a large magnitude, requiring use of very large and very small numbers. To overcome the inconvenience of handling large

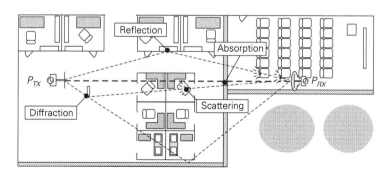

FIGURE 2.7   Indoor radio wave propagation effects

numbers, engineers use decibel (dB) to describe signal levels. A decibel is one tenth of a Bel. A Bel is a ratio of two power levels and is named in honor of Alexander Graham Bell; that's why the B is capitalized. Bel and decibel are defined as follows:

$$Bel = \log_{10}\left(\frac{P1}{P2}\right); \quad dB = 10 \cdot \log_{10}\left(\frac{P1}{P2}\right)$$

P1 and P2 are power levels with the same units.

dB is a logarithmic measure, and it produces easy to handle numbers for large-scale variations in signals. It is useful because system gains and losses can be calculated by adding and subtracting numbers. Remember high school algebra: $log\ (a \times b) = log\ a + log\ b$ and $log\ (a/b) = log\ a - log\ b$. Like percentage, a ratio of two quantities, dB has no units. Let's consider antenna gain and cable losses, both of which are defined later in this chapter. Gain, a ratio of output to input where output is greater than input, is always greater than 1 and thus has a positive dB value. Loss, again a ratio of output to input but where output is less than input, has a value less than 1 and thus has a negative dB value. The log of numbers greater than 1 is positive and for numbers less than 1 it is negative.

To make this concept clearer, here are some examples of dB usage:

- Gain has positive value while loss has negative value.
- Every time you double (or halve) the power level, you add (or subtract) 3 dB to (or from) the power level.
- A 10 dB gain/loss corresponds to a ten-fold increase/decrease in signal level.
- A 20 dB gain/loss corresponds to a hundred-fold increase/decrease in signal level.
- If a cable has 2 dB loss (a typical value for the most antenna cables), it will lose 37 percent of its signal by the time it gets to the other end.
- If a cable has 20 dB loss, it will lose 99 percent of its signal by the time it gets to the other end.

> ### Exam Tip
> Know that a gain of 3 dB is 2 times, 10 dB is 10 times, and 20 dB is 100 times. Gains have positive values while losses have negative values.

The power of a radio signal is expressed in watts. Since the power levels vary by a large amount, it is easier to express them in dBm, decibel relative to 1 milliwatt, rather than in watts. The relation between dBm and watts is defined as follows:

$$Power(dBm) = 10 \cdot \log_{10}\left(\frac{Power}{1mW}\right)$$

The following table shows power values in watts and their equivalent values in dBm.

Watt	10.0	4.0	1.0	0.1	0.01	0.001	10^{-6}	10^{-12}
dBm	40	36	30	20	10	0	-30	-90

Exam Tip
When performing calculations during the exam, remember this rule: 30 dBm = 1 watt; adding 3 dB doubles the power, while subtracting 3 dB halves the power.

Modulation and Encoding

The data, consisting of ones and zeroes, communicated between tags and readers must be sent in a reliable manner. Two steps are critical to reliable communication: the encoding of the data and the transmission of the encoded data—that is, the modulation of the communication signal. The combination of coding and modulation schemes determines the bandwidth, integrity, and tag power consumption.

The prime purpose of radio communication is to convey information from one place to another through the intervening media without wires—for example, an RFID interrogator sending or receiving data to or from an RFID tag. A carrier wave, typically a sinusoidal wave, is used to transport the information. The carrier wave is a radio wave of a specific frequency and amplitude broadcast by the transmitter to carry information. Information is conveyed by modulating the carrier wave. The three key parameters of a wave are its *amplitude*, its *phase*, and its *frequency*, all of which can be modified in accordance with an information signal to obtain the modulated signal. A device that performs modulation is known as a *modulator* and a device that performs the inverse operation of demodulation, to extract information from the modulated wave, is known as a *demodulator*. The actual information in a modulated signal is contained in its sidebands, or frequencies added to the carrier wave, rather than in the carrier wave itself. The most common types of modulation are *amplitude modulation* (AM), *frequency modulation* (FM), and *phase modulation* (PM).

AM varies the intensity, or amplitude, of the carrier wave. When the carrier is thus modulated, a fraction of the power is converted to sidebands extending above and below the carrier frequency by an amount equal to the highest modulating frequency. The modulating signal is recovered by filtering out carrier frequency. Since the carrier wave does not contain any information but is sent along with information, the power used for carrier wave is wasted. There are many variants of AM, such as single sideband modulation (SSB) to increase transmission efficiency. FM varies the frequency of the carrier wave in such a way that the change in frequency at any instant is proportional to the information signal that varies with time. PM changes the phase of the carrier wave. The two methods, FM and PM, are similar in the sense that any attempt to shift the frequency or phase is accompanied by a change in the other.

The modulations AM, FM, and PM are sometimes known as *continuous wave modulation methods* to distinguish them from *pulse code modulation* (PCM), which is used to encode both digital and analog information in a binary way. Other popular binary modulation schemes are amplitude shift keying (ASK), frequency shift keying (FSK), phase shift keying (PSK), and quadrature amplitude modulation (QAM).

To transport digital bits of data across carrier waves, encoding techniques have been developed, each with its own pros and cons. Modulation can be considered a special case of encoding, though the terms tend to overlap in ordinary usage. Technically speaking, modulation involves combining two signals, either of which can be analog or digital, to produce a resultant signal, which can be analog or digital. Encoding, then, is the representation of data by a signal using any method.

Some of the encoding techniques are non-return-to-zero (NRZ), return-to-zero (RZ), Manchester Phase Encoding (MPE), pulse interval encoding (PIE), and FM0. Some of these encoding schemes are discussed when RFID protocols that use them are covered later in the book. Figure 2-8 shows an example of modulation and encoding schemes.

Inductive and Backscatter Coupling

Passive tags do not have an active transmitter that communicates with the interrogator. They typically couple the transmitter to the receiver with either load modulation or backscatter, depending on whether the tags are operating in the

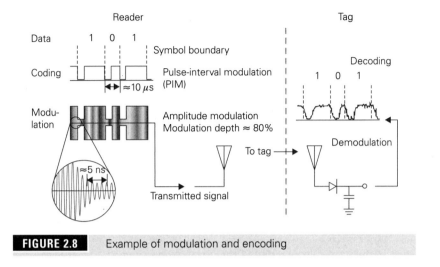

FIGURE 2.8   Example of modulation and encoding

near-field or far-field of the interrogator. *Coupling* is the transfer of energy from one medium to another medium, and passive tags use it to obtain power and transfer data. The type of coupling used, inductive or backscatter (also known as *radiative*), depends on the frequency and the distance between the tag and the interrogator antenna. *Inductive* coupling uses near-field effects, while backscatter coupling uses far-field effects. Near- and far-field effects are different mechanisms for transfer of energy in free space.

The space surrounding any antenna can be divided into two parts, depending on the frequency and antenna size. The primary magnetic field begins at the antenna and induces electric field lines in space. This area is termed the *near-field* of an antenna. The zone where the electromagnetic field separates from the antenna and propagates into free space as a plane wave is termed the *far-field*. In far-field, the ratio of electric field (E) to magnetic field (H) has the constant value of 120ð, or 377Ω. The boundary between near-field and far-field depends on frequency of transmission and size of transmitting antenna. Because changes in electromagnetic fields occur gradually, this boundary is not exactly defined. Several different formulas are used to estimate it. Since the equations used to calculate this boundary are very complex, two simplified approximate formulas are generally used. For transmitting antennas that are electrically small compared to a wavelength (ë) of transmission, the approximate outer edge (r) of the near-field is given by

$$r = \lambda / 2\pi$$

For transmitting antennas that are electrically large compared to a wavelength (ë) of transmission, the approximate outer edge (r) of the near-field is given by

$$r = 2D^2 / \lambda \text{ (where D is the largest dimension of the antenna)}$$

In the near-field, the 120π relationship between (E) and (H) is not necessary. It is possible to have electric field with very little magnetic field, or vice versa. The design of interrogator antennas dictates the choice between these two alternatives. RFID systems are generally designed to minimize any incidental electric field generation. Within the near-field, the magnetic field strength attenuates according to the relationship $1/d^3$; the magnetic field intensity decays rapidly as the inverse cube of the distance (d) between the interrogator antenna and the tag. When the magnetic field strength is translated into power available to the tag, the power attenuates according to $1/d^6$. The magnetic field strength is thus high in the immediate vicinity of the transmitting antenna, but its level is reduced to being negligible in the far-field. This helps create a spatially well-confined IZ filled with magnetic field. In the near-field, the energy field fluctuates outward from the radiating element and then back in to the radiating element in a "push–pull" manner. Beyond the near-field, the electromagnetic energy simply radiates outward and the power drops off based on the inverse-square law. Beyond the near-field boundary, the dominant fields are those associated with energy propagation by electromagnetic waves away from the sources. This region is known as the *far-field region*. In the far-field region, the energy is continuously transported away from the transmitting antenna.

RFID systems operating at 125–135 kHz and 13.56 MHz operate in the near-field and use inductive coupling, while those operating beyond 100 MHz, such as 860–960 MHz and 2400 and 5800 MHz, operate in the far-field and use backscatter (radiative) coupling.

Near-field/Inductive Coupling

Inductive coupling refers to the transfer of energy from one circuit component to another through a shared magnetic field. A change in current flow through one device induces current flow in the other device. Coupling may be intentional or unintentional. Unintentional coupling, called *cross-talk*, is a form of interference. Inductive coupling favors low-frequency energy sources.

The wavelength of the frequency ranges used in inductively coupled RFID systems (135 kHz: 2400 m; 13.56 MHz: 22.1 m) is much more than the conductor length in any standard interrogator antenna. It is also several times greater than the distance between the interrogator antenna and the tag antenna. Because of this, the electromagnetic field may be treated as a simple magnetic alternating field with regard to the distance between tag and antenna. Since the operating wavelength is very much greater than the distance between the tag and interrogator, the coupling elements are not really antennas in the true sense, because no real power is being radiated. Power transfer to and from the tag is by coupling of reactive near-field energy in the immediate vicinity of the antenna structure.

The interrogator antenna, typically a coil, generates a strong, high-frequency electromagnetic field. A small part of this field penetrates the tag antenna, also typically a coil that is a small distance away from the interrogator antenna. By induction, a voltage is generated in the tag antenna, just as in ordinary transformers. This voltage is rectified and provides power supply for the integrated circuit (IC) on the tag. A capacitor is connected in parallel with the interrogator antenna. Its capacitance is selected such that it combines with the inductance of the antenna to form a parallel resonant circuit with a resonant frequency that corresponds to the transmission frequency of the interrogator. Very high currents are generated in the interrogator antenna by resonance step-up in the parallel resonant circuit, which can be used to generate the required field strengths for the operation of the remote tag. This shows that the inductively coupled systems are based upon a transformer-type coupling between the primary coil in the interrogator and the secondary coil in the tag. The optimum number of turns needed in an antenna coil is inversely proportional to operating frequency. As the frequency of operation increases, the number of turns required reduces.

The interrogator communicates with the tag by modulating a carrier wave by varying the amplitude, phase, or frequency of the carrier. This modulation can be directly detected as current changes in the coil of the tag. The tag communicates with the interrogator by varying how much it loads its antenna. This in turn affects the voltage across the interrogator's antenna. By switching the load on and off rapidly, the tag can create sideband frequencies, which are then coupled into the interrogator antenna.

For an RFID system operating at 13.56 MHz, the approximate distance at which the near-field zone ends is $\lambda/2\pi$, or 3.5 meters. Beyond this distance, the magnetic field is reduced so low that tags cannot be powered. Due to this, the typical read range for 13.56 MHz tags is less than 1 meter. The read distance depends on the size of the interrogator antenna. Typical handheld interrogators can read an electronic identity from about 2 inches. An interrogator with a large antenna can read a tag up to 2 feet away. Inductively coupled systems are not affected by water or human tissue but are sensitive to metal (conducting material) in the operating zone. Also, as the magnetic field has vector characteristics, the tag orientation influences performance of the system. Since inductive RFID systems are operated in the near-field, interference from adjacent systems is lower compared to radiatively coupled systems.

Any change in the magnetic environment of a coil of wire will cause a voltage to be "induced" in the coil. No matter how the change is produced, the voltage will be generated. The change could be produced by changing the magnetic field strength, moving a magnet toward or away from the coil, moving the coil into or out of the magnetic field, or rotating the coil relative to the magnet. This means that this type of RFID system gets interference from magnets in the IZ. Figure 2-9 shows near-field coupling phenomenon.

Exam Tip

Know that inductive or near-field coupling is used by RFID systems operating at the LF and HF frequency ranges.

Far-field/Backscatter Coupling

As the electromagnetic wave travels away from the antenna, the magnetic field strength rapidly decays and its effect is negligible in far-field regions. Outside the radius of the near-field, the wave from the interrogator propagates outward, never giving energy back to the radiating element. This electromagnetic wave travels outward and encounters the antenna element in the tag. When electromagnetic waves encounter an object that has a dimension of half the wavelength or more, it gets reflected. The efficiency with which an object reflects electro-

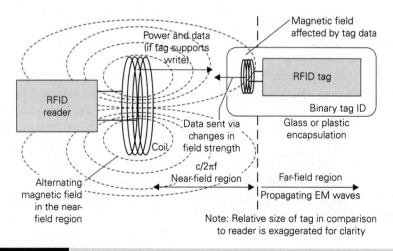

FIGURE 2.9   Near-field coupling (courtesy of Intel Corporation)

magnetic waves is described by its *reflection cross-section*. Objects that are in resonance with the wave front that hits them, as is the case for an antenna at the appropriate frequency, for example, have a particularly large reflection cross-section.

An electromagnetic field propagates outward from the interrogator's antenna, and a small proportion of that field (reduced by free-space attenuation) reaches the tag's antenna. The power is supplied to the antenna connections as high-frequency voltage, and after rectification by diodes it can be used to power the tag or activate or deactivate the tag. Some proportion of the incoming RF energy is reflected by the antenna and reradiated outward into free space. The amount of energy reflected depends on how well the antenna couples to the electromagnetic wave. RFID tags that use backscatter to reply to their interrogators have antennas that are designed to resonate well with the carrier signal emitted by the interrogator. The reflection characteristics of the antenna, its effective cross-section, can be influenced by altering the load connected to the antenna. To transmit data from the tag to the interrogator, a load resistor connected in parallel with the antenna is switched on and off in time with the data stream to be transmitted. By changing resonant properties of its antenna, the tag makes itself a good or poor reflector. This varies the strength of the signal reflected from the tag, creating a pattern that is detected at the interrogator as data. This technique is referred to as *modulated backscatter*.

Before the backscattered signal arrives at the interrogator antenna, it goes through forward and backward path loss, many types of interferences in both the directions, and absorption by the tag to power it. The reflected signal also travels into the antenna connection of the interrogator in the reverse direction from the original signal. It is decoupled using a directional coupler and transferred to the receiver input of the interrogator. The forward signal of the transmitter is to a large degree suppressed by the directional coupler.

Tags operating at UHF and microwave frequencies use far-field and couple with the interrogator using backscatter. The amount of energy received at the receiver decreases as an inverse of the square of the distance (r) between the interrogator antenna and tag ($1/r^2$). Figure 2-10 shows far-field coupling phenomenon.

Exam Tip

Remember that radiative, backscatter, or far-field coupling is used by RFID systems operating at UHF and microwave frequency ranges.

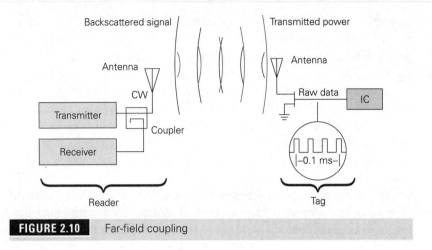

FIGURE 2.10	Far-field coupling

Link Budgets and Read Range

The term *link budget* is used to specify how much loss (attenuation) in signal strength can be tolerated and still allow the receiver to interpret an accurate signal. It is the accounting of all of the gains and losses from the transmitter, through the medium, to the receiver in a radio signal system. It takes into account the attenuation of the transmitted signal due to propagation, as well as the loss, or gain, due to the antenna and cable. Random attenuations such as fading are not taken into account in link budget calculations. Link budget defines the amount of power available in a radio link for transmission loss through the path. With a known link budget, the range of a radio link can be determined given a fixed path loss.

The first step in performing the link budget is determining the required signal strength at the receiver input. This is referred to as *receiver sensitivity* (P_{RX}). A simple link budget equation looks like this:

$$\text{Link Budget} = \text{Transmit Power } (P_{TX}) + \text{Transmit Antenna Gain } (G_{TX}) -$$
$$\text{Cable Loss } (L_C) + \text{Receive Antenna Gain } (G_{RX}) - \text{Receiver Sensitivity } (P_{RX})$$

Because of the large variation in path loss models for propagation in real-world environments, the link budget becomes a more easily comparable specification for evaluation of communication systems. It is generally true that a higher link budget will provide longer range. For this reason, link budget is an important specification for all RF systems.

We will discuss two different link budgets—the forward link and the combined forward and return link budgets—because the requirements for forward link are different from those of return link. The forward link, from interrogator to tag, needs to power the passive tag and also transmit data, while the return link, from tag to interrogator, needs only to transmit data. So we will calculate the forward link budget and compute the maximum range at which a passive tag can be powered. The passive tags do not have their own power source and depend on the power received from the radio waves from the interrogator.

Travel Assistance
For more about passive tags, see Chapter 3.

The combined forward and return link budget will help calculate the maximum distance between the interrogator antenna and the tag at which the interrogator can decode the signals reflected from the tag, also called the *read range*. By comparing these two distances, you can determine the limiting factor for obtaining the maximum read range.

You can calculate the maximum distance at which a passive tag can be powered. A passive tag must receive a certain minimum amount of power to operate. The amount depends on the type of processes used in manufacturing the integrated circuit on the tag. A typical power requirement for a commonly available tag is 100 μW (−10 dBm). Typical power from an interrogator, without considering antenna gain, is 1 W (30 dBm). This means you can sustain a 40 dB free path loss from interrogator to tag, ignoring any interference. A typical passive tag operating in the UHF frequency range has an antenna with an effective collection area of approximately λ/3.5 per side. For simplicity, we use the 900 MHz frequency in this example. You now have all the parameters, so you can calculate the distance at which the tag can be powered:

$$Tag\ area = A = \left(\frac{\lambda}{3.5}\right)^2 = \left(\frac{300/900}{3.5}\right)^2 \approx 0.009\ sq.\ meter$$

$$Path\ loss = -40\ dB = 10^{-4} = \left(\frac{A}{4\pi r^2}\right) = \left(\frac{0.009}{4\pi r^2}\right)$$

$$r = \sqrt{\frac{0.009}{4\pi\left(10^{-4}\right)}} \approx 2.7\ meters \approx 9\ feet$$

According to the assumptions used for the calculations, 9 feet is the maximum distance at which a tag can be powered. Beyond this distance, the tag will not collect enough power to run its IC and thus it cannot be detected by the interrogator. You can increase this distance by employing a higher gain interrogator antenna, increasing the tag antenna effective area, reducing the power requirements of the tag IC, or some combination of these solutions. Typical range for this type of tag is about 15 to 16 feet (4 to 5 meters) when used with a 6 dB gain interrogator antenna.

You can now calculate the maximum range at which a reader can detect the tag if it is not restricted by the requirements of the power. You again start with 1 W power (P_{TX}) from the interrogator and a tag antenna collection area (A_{RX}) of 0.009 square meter. Assume that the interrogator antenna's effective collection area (A_{SC}) is 0.009 square meter. The typical interrogator in UHF frequency using On-Off Keying (OOK) encoding requires a signal-to-noise ratio of 12 dB to detect data (interrogator sensitivity). The typical data rate for this system is approximately 100 Kbps, giving the bandwidth of the signal at around 300 kHz. Now you calculate the noise floor, power of noise, at 300 kHz. The noise power into a 50 ohm resistor at room temperature is −174 dBm/Hz. For 300 kHz, it is −174 dBm × 300,000 (approximately 55 dB).

Noise floor at the signal frequency of 300 kHz = −174 dBm + 55 dB = −119 dBm.

The signal power required = Noise floor + Interrogator sensitivity = −119 dBm + 12 dB = −107 dBm (2×10^{-14}W). This is the minimum power (P_{RX}) that must be received by the interrogator from the tag. You now have all the values you need to calculate the maximum read range. You will use free space path loss from the interrogator to the tag and then back from the tag to the interrogator.

$$\text{Power received by tag} = P_{SC} = P_{TX}\frac{A_{SC}}{4\pi r^2}$$

$$\text{Power received by interrogator} = P_{RX} = P_{sc}\frac{A_{RX}}{4\pi r^2} = P_{TX}\frac{A_{SC}}{4\pi r^2}\frac{A_{RX}}{4\pi r^2} = P_{TX}\frac{A_{SC}A_{RX}}{16\pi^2 r^4}$$

$$P_{RX} = 1\frac{(0.009)(0.009)}{157.8\times r^4} \geq 2\times10^{-14}\ W \ \text{ solve for } r \leq \sqrt[4]{\frac{(0.009)^2}{157.8\times(2\times10^{-14})}} \approx 71\ meters \approx 233\ feet$$

These two calculations show that the link budget from the interrogator to the tag to power the tag is the limiting factor for the UHF tags. If tags can be powered at a greater distance, they can be read further than 16 feet (5 meters).

Exam Tip

For passive RFID tags at UHF and microwave frequencies, know that the read range is limited by the distance at which the tag can be powered, not from how far the tag may be interrogated.

Unlicensed Operation and Frequency Hopping

Unlicensed operation means you are not required to get a permit from a regulatory agency, such as the FCC in the United States, to operate a radio frequency device. Ranges of frequencies have been set aside by regulatory agencies all over the world, and these fall into an unlicensed category collectively called Industrial, Scientific, and Medical (ISM) bands. *Unlicensed* does not mean *unregulated*, however. Many aspects of this band are regulated, including which frequencies can be used for what purposes, maximum amount power that can be radiated, and the duty cycle of the device operation. These regulations are published by the regulatory authority in each country and vary from country to country. All device manufacturers using ISM bands must have regulatory agency certification that their devices operate according to the regulations and must prominently display this certification logo. Unlicensed devices must not cause harmful interference to other radio services. The operator of the device is responsible for correcting any harmful interference that results from the use of the device.

Travel Assistance

Unlicensed operation is covered in more detail in Chapter 10.

Because of their unlicensed nature, ISM bands are popular for establishing wireless data links with short-range wireless devices. With the increase in popularity there is an increase in the number of devices operating in ISM. This has created a congested frequency spectrum, resulting in more interference that can degrade performance of wireless devices.

To reduce the probability of interference, systems are often designed to use *spread-spectrum*. In spread-spectrum, the device uses a narrow band carrier and changes frequency in a given pattern. It transmits at one frequency for a short period of time and then switches to another. Only a fixed number of frequencies are available for a particular group of applications, and they are called *channels*. The switch from one channel to another is called a *frequency hop*.

In a frequency hopping system, the available channel bandwidth is subdivided into a large number of contiguous frequency slots. In any signaling interval, the transmitted signal occupies one of the available frequency slots. Selecting the frequency slots during each signaling interval is made pseudorandomly. Due to the random nature of frequency slot selection, the chances of two similar devices operating at the same slot are reduced. Another advantage of this system is that they are often more immune to interference from other types of systems operating in the same frequency range.

The probability of interception is reduced, because frequency hopping signals have a low average power density at any given slot, which can make these signals difficult to intercept. Also pseudorandom frequency hopping makes these signals difficult to demodulate, if intercepted, by unintended receivers.

Signal Interception and Security

RFID communication occurs in an open environment without any solid encryption. Therefore, the signals, if received by an unintended receiver, can be easily interpreted. The response from tags operating at around 900 MHz can be decoded at up to 71 meters in open air, as was calculated earlier in the discussion of link budget. If this signal passes through a wall, it loses some strength, depending on the type of wall, but it may still be intercepted and demodulated by a good receiver at 20 to 30 meters. The unintended receiver will not be able to power the tag at that distance, but once powered by the authorized interrogator, the signals could be intercepted at a longer distance. In addition to the security issues, problems of reading out-of-process tags, the tags that are not inside the intended read zone, could result. This should be considered when you are designing an RFID system.

Exam Tip

Remember that once powered, passive UHF tags can be interrogated from a distance farther than the normally quoted read range of up to 16 feet.

Antenna Performance and Characteristics

An antenna is a conductive structure that radiates an electromagnetic wave when a time-varying (alternating) electrical current is applied to it. Any similar structure can radiate an electromagnetic wave, but it will not do so efficiently. To radiate efficiently and in a desired manner, the structure must be designed properly as an antenna. In other words, an antenna is a particular arrangement of conductors designed to radiate (transmit) an electromagnetic field in response to an applied alternating electric current. It also generates voltage between its terminals when placed into a time-varying electromagnetic field. An antenna is an electronic component designed to transmit or receive radio waves. It converts electrical energy into a radiating field that extends infinitely outward. All RFID systems include two different types of antennas: the interrogator antenna and the tag antenna. The antenna performance characteristic is one of the most critical elements of any RFID installation, because the antenna transfers power and data from the interrogator to the passive RFID tags and receives the tag's reply.

Design Characteristics

An antenna's performance is affected by several parameters that can be manipulated during antenna design to achieve certain performance characteristics: resonant frequency, impedance, gain, radiation pattern, polarization, efficiency, and bandwidth. Transmit antennas may also have a maximum power rating, and receive antennas differ in their noise rejection properties. An antenna without an amplifier is called a *passive* antenna. Such antennas have the same characteristics whether they are transmitting or receiving. This property is called *reciprocity*. Therefore, any antenna can be used for transmitting or receiving signals or for both purposes. All antennas used in RFID systems are of the passive type.

When the circuit is much shorter than the wavelength of the signal, the rate at which it radiates energy is proportional to the size of the current, the length of the circuit, and the frequency of the alternations. In most circuits, the product of these three quantities is small enough that not much energy is radiated, and the result is that the reactive field dominates the radiating field. When the length of the antenna approaches the wavelength of the signal, the current along the antenna is no longer uniform and the calculation of power output becomes more complex.

Resonant Frequency and Bandwidth

As mentioned, any circuit driven with a signal of high enough frequency will radiate but may not be efficient. Antennas are designed to radiate efficiently at a particular frequency. For example, a dipole antenna (an antenna with two wires connected to a transmission line) that is much smaller than the wavelength of the signal is an inefficient antenna but radiates over a wide range of frequencies. Compare this with a half-wave dipole antenna; it radiates with high efficiency when fed with the signal whose wavelength is twice the electrical length of the antenna. This frequency is called the *resonant* frequency of the antenna. At resonant frequency, as shown in Figure 2-11, the capacitive and inductive components of the antenna impedance become zero.

Most antennas are resonant devices that operate efficiently over a relatively narrow frequency band. Typically an antenna is tuned for a specific resonant frequency and is effective for a range of frequencies usually centered on that resonant frequency. As a tag moves away from this frequency in either direction, the efficiency of the antenna decreases. The resonant frequency is related to the electrical length of the antenna. The physical length of a simple antenna is usually 95 percent of the electrical length. In a complex antenna, the electrical length can be made significantly shorter or longer than the physical length by the addition of an appropriate reactive element. In RFID systems, antennas are

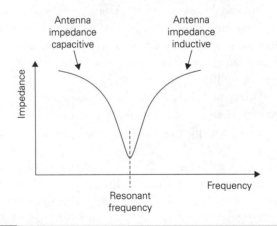

FIGURE 2.11   Resonant frequency

designed for a particular frequency—for example, HF, UHF, or microwave. No antenna tuning is required during installation of the antenna.

The *bandwidth* of an antenna is the range of frequencies over which the antenna is effective. The bandwidth is typically centered on the resonant frequency of the antenna. The UHF RFID systems work at 902 to 928 MHz in the United States. The central frequency of this range is 915 MHz. All the antennas for this system are designed for 915 MHz resonant frequency and they must work well over a band of ± 13 MHz from 915 MHz. Typically, usable bandwidth is between the frequencies at which the *voltage standing wave ratio (VSWR),* a term to be defined in the next section, is less than 2.0. This is shown in Figure 2-12.

Exam Tip	
Know that an antenna transmits a radio signal most efficiently when the antenna resonant frequency matches the signal carrier frequency.	

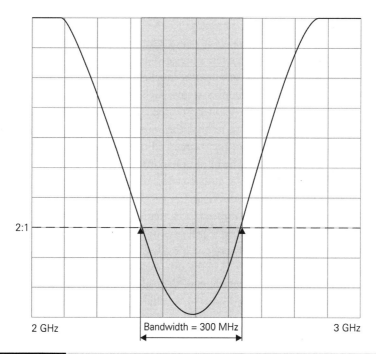

FIGURE 2.12   Antenna bandwidth

Impedance and VSWR

The impedance of an antenna is the sum of its radiation resistance and its ohmic resistance and any reactive components. Power absorbed by the radiation resistance results in power radiated into space, and power absorbed by the ohmic resistance is converted to heat and wasted. Radiation resistance varies with antenna length. Resistance increases as the electrical length of the antenna increases. An antenna circuit can be analyzed as a simple series circuit with resistive, inductive, and capacitive (RLC) reactance. All of these quantities vary with the length of the antenna. For an antenna much smaller than half the wavelength of the signal, resistance and inductance are very small and capacitance is quite large. As the antenna length increases, resistance and inductance increase while the capacitance decreases. An antenna size equal to approximately a half wavelength, capacitance and inductance are equal and cancel each other out, and the resistance is called antenna *impedance*.

One of the goals of antenna design is to minimize the reactance of the device so that it appears as a pure resistive load. Reactance causes energy to be wasted as heat. Reactance can be eliminated by operating the antenna at its resonant frequency, where its capacitive and inductive reactances are equal and opposite, resulting in a net zero reactive current. Once the reactance has been eliminated, what remains is a pure resistance, which is the sum of two parts: the ohmic resistance of the conductors and the radiation resistance. The greater the ratio of radiation resistance to ohmic resistance, the more efficient the antenna becomes.

Different parts of the antenna systems, such as interrogator, feed line, antenna, and free space, may have different impedance. At each interface between differing impedance, some fraction of the wave's energy reflects back to the source. This reflection back to the source and the original wave traveling forward form a standing wave. The ratio of maximum power to minimum power in the wave can be measured and is called the *voltage standing wave ratio (VSWR)*. The ideal value for VSWR is 1:1, which corresponds to a perfect impedance match. Minimizing impedance differences at each interface (impedance matching) will reduce VSWR and maximize power transfer through each part of the antenna system. Therefore, VSWR is a measure of impedance mismatch between the two interfaces in the antenna system. Almost all radio equipment is built for an impedance of 50 ohm. The higher the VSWR, the greater the mismatch and lesser the power radiated. A value of 2:1 VSWR, which is equal to 90 percent power absorption, is considered very good for a small antenna; 3:1 is considered acceptable, and is equal to 75 percent power absorption. Impedance mismatch can occur due to bad connections, incorrect antenna length, or wrong feed line cable.

> ### Exam Tip
> Remember that RFID systems use antenna and cable with 50 ohm impedance. The ideal value for VSWR is 1:1 and as this value increases, the efficiency decreases.

Gain, Directivity, and Radiation Patterns

Before getting into antenna gain, you need to understand a few terms used in describing it. A *dipole* antenna is an antenna with a feeder connected at the center and two wires spreading away from the feeder. An *isotropic* antenna is a theoretical antenna that radiates equally well in all directions. This is a conceptual antenna used for comparison purpose only. An *omnidirectional* antenna radiates equally in all directions in one plane. You will learn more about these in the section "Types of Antennas" later in this chapter.

Gain is the ratio of the signals, usually expressed in dB, received or transmitted by a given antenna as compared to an isotropic or dipole antenna. Antenna gain can be achieved only by making an antenna directional—that is, by having better performance in one direction at the expense of the other directions. The left side of Figure 2-13 shows the antenna gain and its radiation pattern. The middle portion shows the equivalent isotropic antenna. The gain of an antenna is a passive phenomenon. Power is not added by the antenna but is simply redistributed to provide more radiated power in a certain direction than would be transmitted by a reference antenna. If an antenna has a positive gain in some directions, it must have a negative gain in other directions (conservation of energy). The gain of an antenna is therefore a tradeoff between the angle of coverage and power in a covered space. The reason gain is necessary is because we may want to radiate only in a specific direction at a longer distance. For example, an antenna communicating with a satellite needs a very narrow beam and a very high gain.

An antenna *aperture* is the diameter of the cross-section of an antenna's radiation pattern in the direction of highest gain. This frequently is assumed to be circular and is provided as a radius or the angle of the radiation cone. The *beamwidth* of an antenna is a measure of the directivity of an antenna and is usually

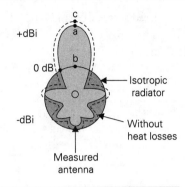

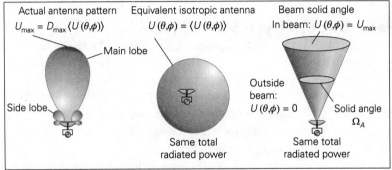

FIGURE 2.13 Antenna gain and radiation pattern

defined by the angle at which the pattern drops to one half of its peak value. This is also referenced as 3 dB beamwidth.

The radiation pattern is the three-dimensional plot of the gain in all directions. Usually only the two-dimensional horizontal and vertical cross-sections of the radiation pattern are considered. Antennas with high gain typically show side lobes in the radiation pattern. Side lobes, shown in the upper-right of Figure 2-13, are peaks in gain other than the main lobe (the "beam"). Side lobes detract from the antenna quality and reduce gain in the main lobe by distributing power in the side lobes that is not useful.

The gain of an antenna is a ratio of two quantities, and thus the reference point of the ratio must be specified. It is expressed in either dBi (decibels over isotropic) or dBd (decibels over dipole) units. If the reference antenna is an isotropic antenna, the gain is expressed in dBi. When a dipole antenna is used as the reference, the gain of the antenna is expressed in dBd. Many times antenna specifications mention only dB rather than dBi or dBd; in this situation, it is usually understood to be dBd, but using an antenna in which the gain is explicitly specified as dBi or dBd can help you avoid surprises later. Gain specified in dBi is always a higher number than gain specified in dBd.

An isotropic antenna cannot be created in the real world, but it is useful for calculating theoretical fade and system operating margins. A dipole antenna has 2.15 dB gain over a 0 dBi isotropic antenna; therefore 0 dBd is equal to 2.15 dBi. So if an antenna gain is provided in dBd, not in dBi, you can add 2.15 to it to get the dBi rating. For example, if an antenna has 6 dBd gain, it would have 6 + 2.15 = 8.15 dBi gain. Antenna manufacturers specify gain in dBd because they calibrate their equipment using a simple dipole antenna as the standard. Then they replace it with the antenna they are testing. The difference in gain (in dB) is in reference to the signal from the dipole antenna, providing gain in dBd.

Two other terms used in radiocommunications are *effective radiated power* (ERP) and *effective isotropic radiated power* (EIRP). These terms are used to describe power output from an antenna. ERP is the product of the power supplied to the antenna and its gain relative to a half-wave dipole antenna in a given direction, usually in the direction of maximum gain (main lobe). EIRP is the product of the power supplied to the antenna and its gain relative to an isotropic antenna in a given direction, usually in the direction of maximum gain (main lobe). The ERP and EIRP are simply the transmitter power times the antenna gain and can be expressed in units of watts, milliwatts, dBw, or dBm. Remember that in dB units, multiplication becomes addition due to the logarithmic nature of dB.

Let's consider two examples of EIRP calculation, one using watts and another using dB. An interrogator connected to an antenna with 6 dBi gain is emitting 1 W of power at the antenna terminal (we will ignore cable losses). What is the EIRP from the antenna?

- **Using Watts**   6 dBi antenna gain means the antenna amplifies the signal by a factor of 4. So EIRP = interrogator power output × antenna gain = 1 × 4 = 4 watts.
- **Using dB**   Converting 1 W from interrogator into dB units gives 30 dBm. Now calculate EIRP = interrogator power output + antenna gain = 30 dBm + 6 dBi = 36 dBm.

Using either method, we get the same answer, because 4 watts = 36 dBm.

In this example we could have used 9 dBi antenna and 0.5 W power from the interrogator and would have reached the same power radiating from the antenna, 36 dBm. But this setup has an advantage over the previous one, because if the same antenna is used to transmit as well as receive, the higher antenna gain will apply to the received signal also. Since the tags receive and backscatter the same power in both the cases, the higher gain would provide higher powered receive signals to the interrogator.

Exam Tip

Remember that the typical value for maximum interrogator power is 30 dBm or 1 watt, for antenna cable loss is 2 dB, and for antenna gain is 8 dBi.

Polarization

Electromagnetic waves, traveling in free space, have an electric field component (E) and a magnetic field component (H). These two components are perpendicular to each other, and both are also perpendicular to the direction of propagation of the wave. The orientation of the (E) vector is used to define the polarization of the wave. If the (E) field is orientated vertically, the wave is said to be vertically polarized. If it is orientated horizontally, the wave is said to be horizontally polarized. When the (E) field rotates with time, the wave is said to be circularly polarized. Polarization of the radiated wave plays an important role when you are concerned with the coupling between two antennas. Figure 2-14 shows polarization of a radio wave.

When electromagnetic waves propagate from an antenna, the position and direction of the (E) field of the wave with reference to the Earth's surface, the ground, determines the polarization of the antenna. The antenna is said to be horizontally polarized if the (E) field is parallel to the ground and vertically polarized if (E) field is perpendicular to the ground. Both of these polarizations are called *linear polarization*, where (E) field always remains in one plane. When (E) field rotates over time, the antenna is said to be *circularly polarized*. If the (E) field rotates in right-hand/left-hand manner, the antenna is considered to be *right-hand/left-hand circularly* polarized. The circularly polarized antenna transfers only half of its total power in any one polarization.

For a receiving antenna to be able to receive radio signals, its polarization must match that of the transmitting antenna. If the polarization of two antennas does not match, lower radiative coupling can take place between them and

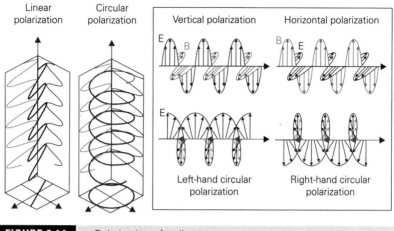

Linear polarization	Circular polarization	Vertical polarization	Horizontal polarization
		Left-hand circular polarization	Right-hand circular polarization

FIGURE 2.14   Polarization of radio wave

less information can be transferred between them. A right-hand circularly polarized antenna cannot couple with left-hand circularly polarized or linearly polarized antenna; they must be matched.

Travel Advisory

If a vendor specifies antenna gain only in dB, be sure to find out whether it is dBi or dBd. If the vendor cannot tell you or does not know the difference, you should consider buying an antenna from another vendor.

Antenna polarization has many implications for RFID systems. A horizontally/vertically polarized antenna will optimally read tags that are oriented horizontally/vertically and will not read as well the tags that are oriented vertically/horizontally. A circularly polarized, right-hand, or left-hand antenna will read tags in any orientation but will provide only half the power of an identical but linearly polarized antenna. Two antennas of different polarization operating in close vicinity will not interfere with each other. Even a right-hand circularly polarized antenna will not interfere with a left-hand circularly polarized antenna. An antenna's polarization is given in its specification, usually along with its gain. For example, an antenna with an 8 dBi gain and linear polarization would be specified as *8 dBil*, where *l* in the specification stands for linear polarization.

This brings us to the four different ways of specifying antenna gain, shown with their meaning here:

- **dBil**   Relative to isotropic antenna with linear polarization
- **dBic**   Relative to isotropic antenna with circular polarization
- **dBdl**   Relative to dipole antenna with linear polarization
- **dBdc**   Relative to dipole antenna with circular polarization

Exam Tip

Remember that antennas with different polarization do not communicate and also do not interfere with each other.

Antenna Performance

Antenna efficiency is the ratio of power actually radiated from the antenna to the power provided to the antenna terminals. Radiation in an antenna is caused by radiation resistance, which can be measured only as a part of the total resistance, including the loss resistance. The loss resistance usually results in heat generation rather than radiation, and therefore reduces efficiency. Impedance of an antenna can be easily measured with specialized equipment. Measuring the radiation pattern requires a sophisticated setup, including an anechoic chamber (to exclude radiation noise) designed for antenna measurements, precise placement of measuring equipment, and specialized equipment that rotates the antenna during the measurements. All of the antenna parameters are expressed in terms of a transmission antenna but are identically applicable to a receiving antenna due to reciprocity. The measurement of impedance at the load, where the power is consumed, is most critical. For a transmitting antenna, the load is the antenna itself, while for a receiving antenna, the load is the receiver rather than the antenna.

The reading range of an RFID antenna depends on many variables: the antenna size, the tag size, the tag's orientation with respect to the transmitting antenna, the antenna location with respect to other materials, and the ambient electrical and magnetic noise within the band of interest. The impedance of an antenna should match that of its connecting transmission line. The polarization of an antenna should match with the tag orientation.

Antennas used for transmission have a maximum specified power rating, beyond which heating, arcing, or sparking may occur in the components, which may cause them to be damaged or destroyed. This, of course, is a concern only for transmitting antennas; the power received by an antenna rarely exceeds the microwatt range.

Types of Antennas

Following are brief descriptions of a few common antenna types found in many RFID systems.

Isotropic Antenna

The isotropic radiator is a purely *theoretical* antenna that radiates equally in all directions. It is considered to be a point in space with no dimensions and no mass. *This antenna cannot physically exist*, but is useful as a theoretical model for comparison with all other antennas. Most antennas' gains are measured with reference to an isotropic radiator and are rated in dBi (decibels with respect to an isotropic radiator). An isotropic antenna's radiation pattern is spherical.

Dipole and Monopole Antennas

The dipole antenna is simply two thin conductors (wires) pointed in opposite directions, arranged either horizontally or vertically, with one end of each conductor connected to the feed line and the other end hanging free in space. Both the conductors usually have the same lengths, and their combined length determine the antenna's wavelength. The two conductors are driven by a voltage applied to the feed line in the middle. The ideal length for a dipole antenna tuned to 915 MHz is about 16 cm (6.5 inches). It is has a modest gain of 2.15 dBi. Since this is the simplest practical antenna, it is also used as a reference model for other antennas. When a dipole is used as a reference to calculate gain of an antenna, the gain of that antenna is expressed as dBd. Variations of the dipole include folded dipole, half-wave antenna, and monopole. Monopole is half the dipole. Monopole antennas are vertical antennas erected on the ground, where the ground, being a conductive material, acts as a reflector, providing an image of the monopole and effectively creating the dipole. Monopole antennas are used for radio transmission, but they are generally less efficient than equivalent dipole antennas. Figure 2-15 shows dipole and monopole antennas.

Generally, the dipole has a toroidal (donut-shaped) radiation pattern, where the axis of the toroid is aligned with the axis of the dipole. This pattern is shown on the left side of Figure 2-16. Most of the power transmitted by a dipole is in horizontal (H) plane with an omnidirectional pattern. In the other two planes,

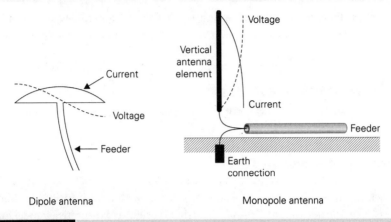

Dipole and monopole antennas

perpendicular to the H plane, it has a transmission pattern like a figure 8. These planes are shown on the right side of Figure 2-16. It has nulls, no power radiated, along the north and the south poles.

Exam Tip

Remember that a dipole antenna has nulls at its north and south poles. Therefore, when a tag antenna pole is pointing in the direction perpendicular to interrogator antenna plane, the tag cannot be interrogated.

Antenna Arrays

When several antenna elements—usually single wire dipoles and reflectors—are combined, they form an antenna array. This provides a particular

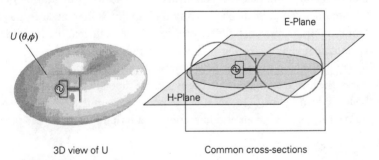

Simple dipole antenna radiation pattern

radiation pattern and gain in a desired direction or is useful when it is difficult to achieve desired electrical properties with a single antenna. Two common antenna arrays are the Yagi-Uda antenna, usually called Yagi, and the patch antenna. The Yagi antenna is sometimes called a beam antenna. This antenna was common before cable and satellite for television and was mounted on roofs for receiving television transmission.

Patch antenna consists of a thin square of conducting material, approximately half wavelength on a side that is closely spaced above a larger reflector plane. The center point of this active element may be grounded using a vertical conductor. A patch antenna can be thought of as a pair of half wavelength slot antennas, spaced by half a wavelength. The antenna impedance varies with the location of the feed point for the active element. The active elements may be of almost any shape, including round and square. A square active element has approximately 0.47 wavelengths on each side, while a round one is 0.54 wavelengths in diameter. The spacing between the active element and the reflector is > 0.01 wavelength, and spacing less than this results in reduced efficiency. For RFID systems, patch antennas are usually designed to have circular polarization and about 6 dBi to 8 dBi gain. Figure 2-17 shows a patch antenna and Figure 2-18 shows a patch antenna array.

Other Antenna Types

Tunnel antennas are used at many conveyor systems. The antenna wraps around the conveyor and allows the reader's energy field to be uniformly radiated inside the tunnel. This allows tags to be read regardless of their orientation or position. A tunnel antenna placed over a conveyor with metal rollers creates a challenging situation for the reader to couple with the tags. Therefore, during the design of the tunnel an-

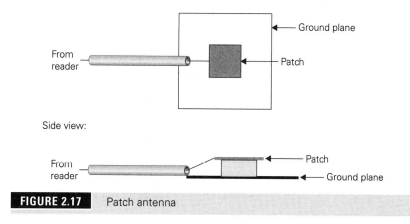

FIGURE 2.17   Patch antenna

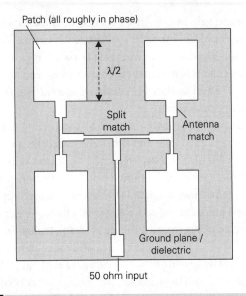

Patch (all roughly in phase)

λ/2

Split match

Antenna match

Ground plane / dielectric

50 ohm input

FIGURE 2.18 Patch antenna array

tenna, special consideration is necessary to prevent the absorption and distortion of the magnetic and radio field energy. Figure 2-19 shows a tunnel antenna.

FIGURE 2.19 Tunnel antenna

Loop antenna consists of a conducting coil leading from one conductor of a two-wire transmission line to the other conductor. The principal lobe of the radiation pattern is wide and is in the direction perpendicular to the plane of the loop. A loop antenna is very directional and its gain is proportional to the loop diameter: the larger the diameter, the higher the gain.

Cables and Connectors

Cables connecting the interrogator with antennas are many times called *transmission lines*. Some cables are permanently connected to the antenna at one end, while others have different types of connectors at both the ends.

Cables come in two different varieties. One type of cable is a balanced twin conductor made up of two parallel conductors. A second type of cable is a coaxial unbalanced line made of two coaxial conductors. In RFID systems, all the cables used to connect antennas with the interrogator are coaxial type. The coaxial cable, shown in Figure 2-20, consists of four parts: the inner conductor, which is either a solid or stranded wire; a surrounding dielectric material; the outer conductor made of braided strands or foil; and the outermost insulator jacket. The outer conductor provides a shield against RF leakage from and to the cable. The effectiveness of the shield depends on the quality and density of the braid. At high frequencies, the braid must be tighter. Coaxial cables come in various diameters, the higher the diameter, the lower the loss but the higher the cost. Thicker cables are difficult to install and have a higher minimum bending radius. The following table shows a few cable types and their attenuation per 100 feet cable length.

Cable Type	Diameter	Loss at 13.56 MHz per 100 feet	Loss at 900 MHz per 100 feet
RG-174	0.100 inch	3.0 dB	26 dB
RG-58	0.200 inch	1.6 dB	15 dB
RG-8X	0.400 inch	1.3 dB	10 dB
LMR-195	0.195 inch		11.1 dB
LMR-240	0.240 inch		7.6 dB
LMR-400	0.400 inch		3.9 dB
LMR-600	0.600 inch		2.5 dB
LMR-900	0.900 inch		1.7 dB

Cables used in RF systems must have impedance that matches the components to which they are connected. Mismatched impedance will reduce efficiency of the system and may create a problem of radiation leakage. All RF devices used in RFID systems, interrogators, and antennas have an impedance of 50 ohms. Therefore, the cables should also have the impedance of 50 ohms.

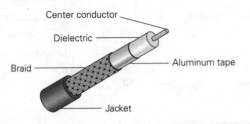

| FIGURE 2.20 | Coaxial cable |

RF connectors are attached to ends of coaxial cables used to interconnect various RF devices. The connectors include threaded or bayonet-style couplings that snap, screw, or push into place and come in many different types. The most common type used in RFID systems is SMA (SubMiniature version A). It uses a threaded interface. SMA connectors of 50 ohms are semi-precision, subminiature units that provide excellent electrical performance from DC to 18 GHz. These high-performance connectors are compact in size and have outstanding mechanical durability. They are intended for use on semi-rigid cables and in components that are connected infrequently (a few hundred interconnect cycles). SMA connectors come in different varieties and are available both in standard and reverse polarity, standard and reverse thread, and standard and reverse gender. The mating connectors must form a proper pair; otherwise, they can be permanently damaged and will need to be replaced.

CHECKPOINT

✔**Objective 2.1: Radio Frequency Basics**   RF waves are part of electromagnetic spectrum and propagate in free space away from the source. They get reflected, diffracted, refracted, and scattered, causing multipath. Several signal modulation and encoding schemes are used to transfer data using radio waves. RFID systems use inductive coupling in near-field and backscatter coupling in far-field to supply power to the tags and communicate with them. The link budget methodology is used to calculate the distance from the interrogator antenna at which a tag can be powered and interrogated. RFID systems operate in unlicensed ISM bands and must follow local government regulations regarding usage of these bands.

✔**Objective 2.2: Antenna Performance and Characteristics** Antennas convert electrical energy to radio energy and radiate this energy in free space. The efficiency with which they radiate energy depends on the type of antennas and their design parameters. These parameters include resonant frequency, bandwidth, and impedance, which are related to the frequency of the signal for which the antenna is designed, while the gain of antenna depends on the antenna design. Several different types of antennas are available, but in RFID systems, the most prevalent are dipole and patch antennas.

Increasing antenna gain increases its beam length but decreases its beamwidth. Different objects and their material type encountered by the RF signals during their travel have a great effect on the power and shape of the signals received by the receiver.

Radiated power from an antenna is related to antenna gain, the type of cable used to connect the antenna with the interrogator, and the power output from the interrogator's antenna terminal. Cables have losses that increase with the length of the cable and the frequency of the signal transmitted and decrease with the increase in diameter of the cable. The power radiated from antenna = interrogator power – cable losses + antenna gain when all are expressed in decibel units.

REVIEW QUESTIONS

1. In a UHF RFID system, interrogators and tags are using 900 MHz frequency. What is the approximate wavelength of the radio waves emitted by the interrogator antenna?

   A. 33 feet
   B. 33 meters
   C. 33 centimeters
   D. 6.5 inches

2. What is the wavelength of a sinusoidal wave?

   A. The distance between two consecutive positive peaks on the wave
   B. The distance traveled by the wave during 1 microsecond
   C. The distance traveled by the wave during 1 millisecond
   D. The distance between the positive peak and the negative peak

3. What is the relationship between the power density (PD) of an electromagnetic wave and the distance (d) from the source of the radiated wave?

   A. Path density (PD) is proportional to square of distance (d^2) – PD á d^2

   B. Path density (PD) is proportional to distance (d) – PD α d

   C. Path density (PD) is inversely proportional to (d) – PD α 1/d

   D. Path density (PD) is inversely proportional to (d^2) – PD α 1/d^2

4. Radio waves are highly reflected from what?

   A. Dielectric material

   B. Conductive material

   C. All material

   D. Water

5. If a cable has 13 dB loss, what percentage of the original signal power will be received at the other end of the cable?

   A. 10

   B. 5

   C. 15

   D. 85

6. Which one of the following statements applies to inductive coupling?

   A. Inductive coupling operates in near-field and uses a magnetic field.

   B. Inductive coupling operates in far-field and uses a magnetic field.

   C. Inductive coupling operates in near-field and uses radio waves.

   D. Inductive coupling operates in far-field and uses radio waves.

7. The backscatter coupling phenomenon is used for communication between the interrogator and tags at which of the following frequency ranges?

   A. Low frequency (LF) and microwave frequencies

   B. Low frequency (LF) and high frequency (HF)

   C. Ultra high frequency (UHF) and microwave frequencies

   D. High frequency (HF) and ultra high frequency (UHF)

8. Which of the following is not considered for calculating link budget?

   A. Transmit antenna gain

   B. Cable losses

C. Receive antenna gain

D. Losses due to fading

9. Power output from an antenna terminal of an interrogator is 27 dBm. What is its value in watts?

 A. −1.0 watt

 B. 2.0 watts

 C. 0.5 watt

 D. 1.0 watt

10. Which of the following statements is true for antenna gain?

 A. The higher the antenna gain, the narrower the main beamwidth.

 B. The higher the antenna gain, the higher the total power radiated by the antenna.

 C. The higher the antenna gain, the wider the main beamwidth.

 D. The higher the antenna gain, the shorter the main beam.

11. In a two-antenna communication system, one antenna is fixed while the other rotates in a random manner around a different axis. The rotating antenna is linearly polarized. What is the best polarization for the fixed antenna?

 A. Vertical

 B. Circular

 C. Linear

 D. Horizontal

12. Which of the following does not affect cable losses?

 A. Minimum bending radius

 B. Cable length

 C. Cable diameter

 D. Frequency of transmitted signal

REVIEW ANSWERS

1. **C** Wave length = speed of light / frequency = 300×10^6 (m/sec) / 900×10^6 = 0.33 m = 33 cm.

2. **A** The wavelength of a sinusoidal wave is the distance between two consecutive positive peaks on the wave or a distance traveled by the wave during one complete cycle.

3. **D** Path density (PD) is inversely proportional to (d^2) – PD á $1/d^2$. As the distance from the source of the radiation increases, the power density decreases as the square of the distance.

4. **B** Conductive material, usually metals, highly reflect radio waves. All materials reflect some portion of the wave.

5. **B** Only 5 percent of the signal power will be received. Here's a simple way to calculate this: 13 dB = 10 dB + 3 dB. At 10 dB you loose 90 percent or receive 10 percent. An additional 3 dB loss will halve this signal, giving 5 percent.

6. **A** Inductive coupling operates in near-field and uses magnetic field to transfer energy between two coupling circuits.

7. **C** Ultra high frequency (UHF) and microwave frequencies use backscatter to communicate between interrogator and tags.

8. **D** Losses due to fading or any such random losses are not considered in link budget calculations.

9. **C** 27 dBm = 0.5 watt, 30 dBm = 1 watt, 27 dBm = 30 dBm – 3 dB. Subtracting 3 dB means halving the power, providing 0.5 watt.

10. **A** The higher the antenna gain, the narrower the main beamwidth. As the antenna gain is increased, the width of the main beam decreases but its length increases.

11. **B** Circular polarization is the best in this situation because it will allow the fixed antenna to couple with the rotating one, regardless of the rotating antenna's orientation.

12. **A** Minimum bending radius does not affect cable losses. It provides a guide to the installer about how sharply the cable may be bent without damaging it. Cable losses are directly proportional to cable length and signal frequency and inversely proportional to cable diameter.

Tag Knowledge

ETA

NEWBIE	SOME EXPERIENCE	EXPERT
6 hours	3 hours	1 hour

RFID tags, along with interrogators, antennas, and peripheral devices, form a physical layer of the RFID system. In Chapter 2, you read about the phenomena of electromagnetic wave propagation and inductive and backscatter coupling. RFID tags use these phenomena to communicate with interrogators. In this chapter, you will read about RFID tag components and subassemblies, types of tags and reasons for using various types of tags, tag performance parameters, and tag selection criteria.

A tag is usually designed to suit a particular purpose, such as mounting on metal or embedding in tires. A tag's behavior changes according to the material to which it is attached and the condition of the surrounding environment. Therefore, selecting an appropriate tag for the object material and properly applying it to the object is the most important factor in the successful deployment of an RFID system.

Tag Components and Construction

Objective 3.1

RFID tags are made of three different components: integrated circuit (IC), antenna, and substrate, as shown in Figure 3-1. A tag manufacturer typically does not make all three components in-house. The IC is typically designed and made by a semiconductor manufacturer, while an antenna is usually designed and made by a tag manufacturer. Tags are available in various sizes, designs, and form factors, and they can be customized for a particular application. Unless you need a very large number of tags, you should try to use stock tags because customized tags can be very expensive.

Integrated Circuit

An IC, also called an electronic circuit, microchip, or chip, is designed and manufactured by a semiconductor manufacturer. Designing ICs requires knowledge and experience in the technology, and their manufacture requires expensive equipment. Therefore, ICs are often designed by one company, manufactured by another, and used by a third company that assembles the components to cre-

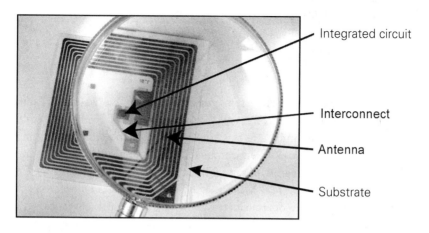

Integrated circuit

Interconnect

Antenna

Substrate

FIGURE 3.1 RFID tag components and interconnection

ate a tag. One IC may be used for many different types of tags by different tag manufacturers.

The IC has a logic unit that makes decisions and provides memory to store data. Actually, the IC is a small microprocessor, and the latest ones have 40,000 to 50,000 transistors, more than the original IBM PC. The IC needs power to operate. This power may come from a battery on the tag (in an active tag) or it may be obtained from the radio energy radiated by the interrogator antenna (in a passive tag). A part of the IC is dedicated to controlling power.

The processing logic implements the communication protocol. It also is used to modulate/demodulate signals and to encode/decode digital bits during the communication between the interrogator and the tag. The memory on the IC may be divided in different blocks, called *banks*. A block may be a read-only type, a write-one-time-only type, or a write-many-times type. Tags use electrically erasable, programmable, read-only memory (EEPROM). This type of memory does not require continuous power to store data. Therefore, data stored on the tag is retained for a long period (several years), even though no power may be available to the tag during this period. The type of data stored in the memory depends on the protocol used. The IC may store tag ID, object identifier, password, and error detection code such as cyclic redundancy check (CRC).

> ## Local Lingo
> **ROM and WORM**   Read-only memory is called *ROM* and write-one-time-only memory is called *WORM*, for write-once-read-many.

ICs are created on a large semiconductor wafer. One wafer can contain 40,000 ICs. Manufacturing of ICs requires state-of-the-art clean room facilities. The finished ICs are individually tested to ensure reliable functionality in the field. With advances in semiconductor technology, ICs are becoming smaller—as small as a grain of sand. The smaller the IC, the lesser the cost and power required to operate it. Making tag ICs more efficient in power usage and requiring less power to operate increases the read range of passive tags.

The ICs on each wafer need to be cut and separated and then attached to a tag antenna. As the size of IC decreases, it requires more precise equipment to connect it to the antenna, which can increase the cost of assembling the tag.

Antenna

This section discusses a type of antenna that is mounted on the tag rather than one connected to an interrogator. The antenna is the largest part of the tag and is connected to the tag IC. The antenna receives the signals from the interrogator and, depending on tag type, it either transmits or reflects the received signal back. For active tags, it transmits the signals, and for semi-passive and passive tags, it reflects the signals. For passive tags, the antenna also collects power from the radio waves and supplies it to the IC.

The geometry of an antenna determines the frequency at which the tag operates. Though the tags may use the identical IC, variations in antenna design allow tags to have completely different properties and behaviors. The antenna can be shaped as a spiral coil, a single dipole, dual dipoles (one perpendicular to other), or a folded dipole. Within these basic types are many variations in antenna shapes, depending on the specific requirements of the application and the abilities of the designer. The antenna is designed for a specific frequency of operation and is later tuned according to the properties of the material to be tagged. The designated frequency determines the effective antenna length, but the actual antenna length is typically reduced using creative antenna design. As mentioned, the antenna is the largest component of the tag, so it determines the physical parameters of the tag.

Antennas are usually made of thin metal strips of copper, aluminum, or silver. These strips are deposited on the substrate at high speeds, and then the IC is attached to the antenna. The antenna can be manufactured using one of the three different methods: copper etching, foil stamping, and screen-printing. Screen-printing is the fastest and the least expensive of all three processes, but the antennas created using this method are less efficient than those created by the other two methods. Because many tags will be included in the labels and the label makers' expertise is in printing, a screen-printed antenna using conductive ink containing copper, nickel, or carbon would make the tag creation process less expensive and will integrate it with the label making process.

Exam Tip
Know that the design of the antenna makes the tags behave differently, though they may use the same IC.

Interconnection

The antenna is connected to the IC on the tag. This connection is usually the tag's "weakest link." The strength of this connection greatly influences the quality and durability of the tag. Even if it is not broken, the connection can be poor and reduce the available power to the tag or reduce the strength of the signal sent back by the tag. Repeated heat cycling or a corrosive environment can degrade the quality of the connection, and poor connection means lower tag performance. The connection may become weak or even break due to repeated bending of the tag during manufacturing of smart labels or during tag application. To avoid damage due to sharp tag bending, tag manufacturers usually specify a minimum bending radius for their tags.

Substrate

The substrate holds all other tag components together. The tag antenna is deposited or printed on the substrate, and the IC is then attached to this antenna. A substrate is usually made from flexible material such as thin plastic, but it may also be made from rigid material. Most passive tags use substrates made from flexible material with a thickness of 100 to 200 μm. The substrate material must be able to withstand various environmental conditions through which the tag may pass during its lifecycle. Some of the materials used for the substrate are polymer, PVC, Polyethylenetherephtalate (PET), phenolics, polyesters, styrene, and even paper. The substrate material must provide dissipation of static buildup, a smooth printing surface for antenna layout, durability and stability under various operating conditions, and mechanical protection for the antenna, chip, and their interconnections. Some of the environmental conditions that can affect the substrate are heat, moisture, vibration, chemicals, sunlight, abrasion, impact, and corrosion. The substrate material may affect the design frequency of the antenna; therefore, the effect of substrate material must be considered during proper tuning of the antenna.

One side of the substrate is usually coated with an adhesive material to attach the tag to an object. The adhesive material must be able to withstand appropriate environmental conditions. Sometimes, a protective overlay made from materials such as PVC lamination, epoxy resin, or adhesive paper is added to protect the tag from environmental effects.

Tag Packaging

Tag components are packaged in various ways to make them easy to handle for tag manufacturers with different levels of expertise in manipulating small electronic components. The manufacturing of the final tags from the various components require organizations with expertise in semiconductor design and manufacturing, RF devices design, manufacturing of printed circuits, and manufacturing and printing of labels. Because of this complexity, tag components are packaged into subassemblies that make tags easier to manufacture. Three commonly used terms in tag manufacturing are *strap*, *inlay*, and *smart label*.

Straps

During tag manufacture, the IC is attached to the antenna. The IC is very small, about 0.25 square mm, and it has two connection points that must be aligned properly and connected to the antenna formed on the substrate. Due to the size of the IC, aligning it with the antenna is difficult and requires precise and expensive equipment. It also requires expertise in manipulating small components at very high speeds. Most tag manufacturers do not have this type of expertise or equipment. To overcome this problem, the IC manufacturers attach one conductive film pad to each connection points of the IC. These pads are larger than the IC and provide a larger area where an antenna may be attached. This subassembly with the conductive film pads and the IC is called a *strap* and is shown in

Figure 3-2. A strap is typically supplied on a reel of continuous web to help the manufacturer attach the antenna to it via high-speed strap attachment equipment. With a larger area of strap pads, the tag manufacturer does not require as much precision equipment and expertise in handling small parts.

Inlays

Another subassembly of tag components is called an *inlay* or insert. This is essentially a complete tag that is usually embedded in a label. An inlay consists of an IC, an antenna, and a continuous length of substrate to accommodate several thousand inlays. The IC used may be part of a strap. Typically, the substrate does not have an adhesive. The inlays are supplied on a reel of continuous web and are used by label makers, also called *converters*, to embed RFID functions into labels. The continuous form of inlays helps assemble labels on high-speed equipment. Figure 3-3 shows a sample of inlay.

Inlays are electronic devices, and just like any other electronic device they may fail due to improper assembly, mechanical damage, or electrostatic discharge (ESD). During inlay and smart label manufacturing, the reels of inlays are wound and unwound at high speeds. These operations may generate an ESD and can severely damage the IC on the inlays. The amount of ESD depends on the winding speed and tension. Manufacturers of inlays recommend using the ESD deionization equipment during high-speed handling of the inlays.

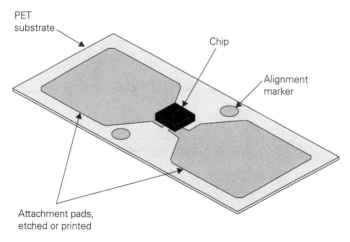

PET substrate

Chip

Alignment marker

Attachment pads, etched or printed

FIGURE 3.2   IC strap subassembly

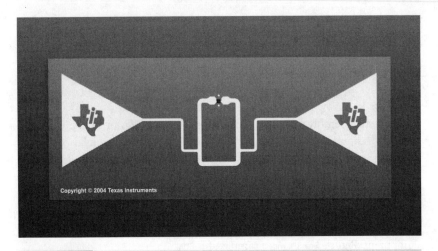

Copyright © 2004 Texas Instruments

FIGURE 3.3   Inlay sample

Smart Labels

Smart labels, often called labels, are made of layers of thin sheets of various types of materials with an RFID inlay embedded within those layers. These layers usually include the linear or carrier sheet, linear release coating, inlay adhesive, tag inlay, face stock adhesive, label material, and label topcoat. Figure 3-4 shows a typical smart label. Barcodes and textual information are usually printed on the smart label material in addition to an identifying number that is encoded into the embedded RFID inlay. The smart labels use either a direct thermal or thermal transfer printing method (see Chapter 5).

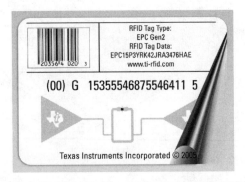

RFID Tag Type:
EPC Gen2
RFID Tag Data:
EPC15P3YRK42JRA3476HAE
www.ti-rfid.com

2035564 020 3

(00) G 15355546875546411 5

Texas Instruments Incorporated © 2005

FIGURE 3.4   Smart label

Smart labels come in various sizes, and depending on the material to be tagged, they have different types of tags embedded into them. They provide logistical information for shipping and handling of the products in the supply chain. They allow RFID encoding to be used while retaining the current barcode and human readable system. Smart labels are available in rolls or in fan folds. They provide an easy way to implement RFID systems within the supply chain, particularly for companies trying to meet the RFID mandates of their customers while simultaneously meeting their other clients' ongoing barcoding requirements.

Smart label stock received from the manufacturer can contain defective labels, though the quality and yield has improved over the years. Smart labels are also affected by ESD generated during transportation, storage, and usage. During encoding and application of the smart label, defective labels must be identified and discarded without disrupting the business processes using the smart labels. Efficient ways of handling defective labels must be integrated into the RFID system during the system design.

Encapsulated Tags

RFID inlays are sometimes encapsulated inside a hard case. The case is made of RF translucent materials such as PET, Polypropylene (PP), Polyacetate (POM), Polycarbonate (PC), Acrylonitrile Butadiene Styrene (ABS), Polyamide 66P (A66), and Elastomère (EPDM). Tags are encapsulated to protect them from harsh environments. For example, a tag attached to a reusable plastic container (RPC) used in a food processing plant would encounter high temperature, pressure, and steam when the container is sanitized. These tags are used to track totes, carriers, and pallets in closed-loop systems and returnable assets in the supply chain. Some of the cases may provide a more rugged method of tag attachment, such as bolting or riveting to the object. In addition, the size and shape of the case may be designed to provide isolation from the object so the tag may be attached to metal objects. Figure 3-5 shows some samples of encapsulated tags.

Local Lingo

RF translucent material   A material that allows RF waves to pass through it with almost negligible attenuation.

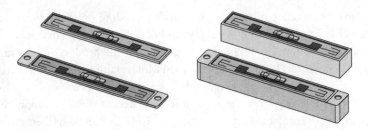

FIGURE 3.5 Encapsulated tags

Exam Tip

Know that encapsulated tags are frequently used for tracking reusable containers in manufacturing and supply chains.

 **Objective 3.2 Tag Types**

RFID tags come in many different designs, shapes, and sizes. A tag is designed for a particular application or a set of applications. Depending on the object or material to which the tag is to be attached, it may require different frequency of operation or functionality. The read range of a tag also varies with the frequency of operation. The tag, being an electronic device, needs power to operate, and tag design dictates the source of this power.

The tag memory may have restriction on how often it can be written to. This may be for security reasons, so that the data, once written, cannot be changed. Some applications may require tags to be of a certain size—for example, a laundry tag embedded inside a garment should be no bigger than a button. A tag embedded inside a tire must be designed to withstand high temperatures and pressure encountered during the manufacturing of the tire. A tag mounted on a metal object requires special design and mounting considerations.

Tags may be classified under four categories, depending on how the tags obtain power, the frequency at which they operate, the protocol used, and the various functionalities implemented on the tags.

Power Source

A tag requires power to process signals received from the interrogator and to send the data encoded signals back to the interrogator. Signals sent back may be reflected signals or signals generated by the tag. Depending on how tags obtain the power and how they use that power, tags are classified as the following:

- Passive tags
- Semi-passive tags
- Active tags

Passive Tags

A passive tag (Figure 3-6) does not have its own power source; it has no battery on-board. The tag obtains power from radio waves received from the interrogator. The amount of power thus received is very small, just enough to energize its IC. Therefore, passive tag functionalities are limited. Due to a lack of enough power, it cannot support an active transmitter to communicate with the interrogator. The good thing about the lack of transmitter, however, is that passive tags do not contribute to radio noise. To communicate with the interrogator, passive tags operating at low and high frequencies use inductive coupling, while those operating above a high frequency range use radiative coupling. (See Chapter 2 for more about these coupling methods.) Inductively coupled tags have a read range of a few inches to about 2 feet, while radiatively coupled tags have read range of up to 20 feet.

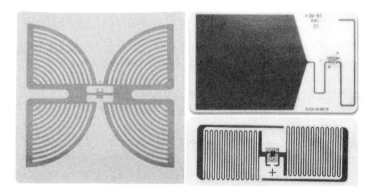

FIGURE 3.6 Passive tags

When a passive tag is not in an interrogation zone (IZ), which is the case most of the time, it has no power and does nothing. Due to this, the passive tag cannot contain any kind of sensors that require continuous power, such as temperature and pressure sensors. The passive tag typically has a minimum functionality of carrying and transmitting a small amount of data. It is a simple and inexpensive device compared to active and semi-passive tags.

In RFID applications, passive RFID tags are often used. Because of their low cost, passive tags are well suited in applications for which tags are not reusable. The tags become part of the object to which they are attached and have the same lifecycle as the object itself; they are not returned into circulation after the object's life has expired. For example, in a supply chain, a case of goods may have passive tags attached to it; when the items are removed from the case and the case is discarded, the tags are also discarded. This makes economic sense due to the low cost of the passive tag.

The following table lists some of advantages and disadvantages of passive tags.

Advantages	Disadvantages
Small size	Requires presence of interrogator to work
Lightweight	Limited amount data storage
Inexpensive (depends on quantity)	Require higher power interrogators
Does not add to radio noise	Low read range (few inches to 20 feet)
Longer life (20-plus years)	
Resistance to harsh environment	

Travel Advisory

A low read range might be considered an advantage or a disadvantage, depending on a tag's application. When a tag is used for a credit card or for an access control card, a low read range is an advantage, because you do not want somebody to be able to read your personal card from 20 feet away. However, when used in a supply chain to tag a pallet, a low read range is a disadvantage because it is difficult and dangerous for a forklift to pass repeatedly very close to the interrogator antenna. Due to the recent hype about using RFID tags in a supply chain, where longer read range is an advantage, and due to the general mentality that bigger is better, people usually consider a low read range a disadvantage. To stay in synch with the RFID industry, this book also assumes that a lower read range is a disadvantage, but it also points out where a longer read range can be a disadvantage.

Semi-Passive Tags

Semi-passive tags (Figure 3-7) are also called semi-active, battery-assisted passive (BAP), or battery-assisted tags (BATs). This tag has an on-board battery to power its IC, but, like a passive tag, it does not have an active transmitter. It uses backscatter to communicate with the interrogator. It modulates the reflection of the waves from the interrogator and requires an interrogator to send data. Since no transmitter is present, it also does not contribute to radio noise. This type of tag is used because it can provide a longer read range than a passive tag and can accommodate environmental sensors on-board. The sensor on the tag helps record the environmental experience of the object to which the tag is attached.

You may wonder why the semi-passive tag includes a battery. First, the battery provides power to the IC so the tag does not depend on the interrogator to power it. If you recall the discussion on link budget in Chapter 2, you'll remember the forward link and return link budgets. The forward link is used to provide power to the tag and to communicate with the tag, and it has a smaller range than the return link. The return link, which includes a round trip from interrogator to tag and back to interrogator, still has a longer range than the forward link. This means that the limiting factor for a read range of a passive tag is how far away from the antenna a tag can be powered, not how far away from the antenna the signals can be decoded. With that in mind, if we provide a battery to power the IC, we can extend the read range of the passive tag. Therefore, by slightly modifying the passive tag to accept the power from on-board battery, we can increase its read range.

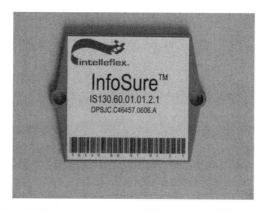

FIGURE 3.7 Semi-passive tags

The second reason to have a battery on the tag is to attach an environmental sensor to the tag. Environmental sensors require reliable continuous power to operate, and they require a higher power level than the tag IC. The passive tag cannot have reliable and continuous power because it gets power only when it is in the IZ, which is not the case most of the time. In addition, when it gets the power inside the IZ, it is very limited—not enough to power an on-board sensor. Therefore, if a battery is added to the tag, the sensors and the tag IC can use that power, which is available all the time. The sensors on the tag can collect data and transmit that data along with the identifying number when the tag is interrogated. This is a valuable prospect, because, for example, it allows you to collect the temperature experience of the object automatically as the object moves around during its lifecycle. Therefore, when a case of temperature-sensitive drugs is transported, you can collect a record of how long the case was subjected to the temperatures beyond certain set limits and dynamically calculate its expiry date or its potency. Frozen food storage conditions can be easily monitored using this type of tag. Temperature, pressure, relative humidity, acceleration, vibration, motion, altitude, and chemical sensors can all be placed on a tag.

The extra functionalities of an on-tag battery do not come without cost, however. The battery creates a few problems, such as extra weight, larger size, higher cost, shorter life, and temperature sensitivity. An integrated battery means the tag dies when the battery dies, and a replaceable battery means larger size and weight for the battery compartment. How do you know when a battery is drained? The interrogator can never detect a tag that does not respond. If the object with the tag is sitting outdoors in below-freezing temperatures, the battery may not work and the tag will not respond. Batteries on tags have a typical maximum shelf life of 2 to 7 years, but when tags are used and interrogated frequently, the batteries can quickly be depleted.

The following table lists some of advantages and disadvantages of semi-passive tags.

Advantages	Disadvantages
Longer read range (100-plus feet)	Requires presence of interrogator to work
Reduced power from interrogators	Larger size and weight
Can have more memory, store more data	Costs a bit more than passive tags
May contain environmental sensors	Limited battery life (2 to 7 years)
Does not add to radio noise	Battery sensitive to harsh environment

You may also ask, "Why not put an active transmitter on semi-passive tags?" An active transmitter requires a lot more electronics on the tags as well as a larger power source. This would make the tags bigger, heavier, and costlier. If you need to acquire data only when the tags are in the IZ, why pay for the extra functionality?

> ### Exam Tip
> Know that tags with batteries on them may not work when they are exposed to subfreezing temperature for an extended period.

Active Tags

An active tag (Figure 3-8) has an on-board power source, usually a battery, and an active transmitter. The IC of this tag may contain more processing power to implement additional functionalities such as data manipulation. This tag uses the battery to power its IC and transmitter. It does not need emitted power or radio signals from the interrogator to transmit its data. Actually, it does not even need an interrogator. An active tag may be set to broadcast its data at a preset time, periodically, or on occurrence of a certain event. Its typical read range is 300 to 750 feet. The read range depends on the battery power and type of transmitter on the tag. An active tag, like a semi-passive tag, may have on-board sensors or external sensors connected to it. With more processing power, the tag may collect data from the sensors and locally process the data before broadcasting. Active tags are often used by real time location systems (RTLSs). (RTLS is discussed in Chapter 5.)

FIGURE 3.8   Active tags

An active tag can communicate with other active tags with or without the presence of an interrogator. It does not communicate with passive or semi-passive tags. An active tag can be designed to broadcast its data using broadband or spread-spectrum technologies to enhance its data communication performance. It can be set to a sleep mode in which it uses only a small amount of power and does not transmit data. When it receives a specific signal, it awakens and transmits its data. This reduces the power usage from the battery and extends battery life. Some active tags work at two different frequencies. At the lower of the two frequencies, they work in receive only mode and use this mode to receive a signal from a nearby transmitter. This signal awakens the tag, and then the tag transmits its data at higher frequency and longer range. During this transmission, it may also transmit the identity of the device that awakened it. This is typically used within a large container yard where you want to track the movement of the container from one zone to another. The tag may also be set to transmit a periodic beacon to let an interrogator know of its existence or send the signal when moved or handled. It can even send a low battery alert. The US Department of Defense has been using this type of tag for many years.

The advantages and disadvantages of active tags are similar to those of semi-passive tags and are listed in the following table:

Advantages	Disadvantages
Longer read range (100-plus feet)	Requires presence of interrogator to work
Reduced power from interrogators	Larger size and weight
Can have more memory, store more data	Costs more
May contain environmental sensors	Limited battery life (2 to 7 years)
Contributes to radio noise	Battery sensitive to harsh environment

---

Exam Tip

Remember that passive and semi-passive tags use backscatter to communicate while active tags use on-board radio to communicate.

Tag Frequencies

RFID tags are categorized according to the frequency at which they are designed to operate. Four primary frequency ranges are allocated by various government authorities for use by RFID systems.

- Low frequency (LF)
- High frequency (HF)
- Ultra high frequency (UHF)
- Microwave frequency (microwave)

All of these ranges are part of the frequency bands called Industrial, Scientific, and Medical (ISM) radio bands and were originally reserved for non-commercial use. Depending on the usage, the tags are designed to operate in any one of these ranges. The material of the object being tagged and the read range required are the main factors in selection of frequency. Objects with water content absorb frequencies in certain ranges but have no effect on frequencies in another range.

Low Frequency (LF) Tags

Low frequency range includes frequencies from 30 to 300 kHz but only 125 kHz and 134 kHz (actually, 134.2 kHz) are used for RFID. This range has been in use for RFID tags for animal tracking since 1979 and is the most mature range in use. It is available for RFID use all over the world. The tags in this range are generally called *LF tags*. They use near-field inductive coupling to obtain power and to communicate with the interrogator. The LF tags are passive tags (no battery and transmitter on the tag) and have a short read range of a few inches. They have the lowest data transfer rate among all the RFID frequencies and usually store a small amount of data. The LF tags have no or limited anti-collision capabilities; therefore, reading multiple tags simultaneously in the IZ is impossible or very difficult. The tag antennas are usually made of a copper coil with hundreds of turns wound around a ferrous core. They are expensive to manufacture, and tags using them are thicker than others at higher frequencies. The LF tags can be easily read while attached to objects containing water, animal tissues, metal, wood, and liquids.

LF tags (Figure 3-9) have the largest installed base. They are used in access control, asset tracking, animal identification, automotive control, as vehicle immobilizers, healthcare, and various point-of-sale applications (such as Mobil/Exxon SpeedPass). The automotive industry is the largest user of LF tags. For example, in an automobile vehicle immobilizer system, an LF tag is embedded inside the ignition key. When that key is used to start the car, an RFID interrogator placed around the key slot reads the tag ID. If the tag ID is correct, the car can be started. If the ID is incorrect or no tag is found, the car cannot be started.

High Frequency (HF) Tags

The high frequency range includes frequencies from 3 to 30 MHz but only one frequency, 13.56 MHz, is used for RFID applications. This frequency is now available for RFID applications worldwide with the same power level. Tags and interrogators using 13.56 MHz are generally called the *HF tags* and the *HF interrogators*. They, like the LF tags, also use near-field inductive coupling to obtain power and to communicate with the interrogator. HF tags are passive tags and have short read range, less than 3 feet. They have a lower data transfer rate than the UHF frequencies but a higher data rate than the LF. The HF tags may have anti-collision capability that facilitates reading of multiple tags simultaneously in the IZ. Since the read range of many HF tags and interrogators is small, they usually do not implement anti-collision. This reduces the complexity and cost.

FIGURE 3.9   LF tags are embedded in these keys

Some HF tags can store up to 4K of data. HF tags are more mature than UHF tags and many standards are in place.

HF tag antennas are usually made of a copper, aluminum, or silver coil with three to seven turns. They are easy to manufacture, and the tags are usually very thin, almost two-dimensional. HF tags are lower in cost due to simpler antenna design than LF tags. They can be made in different sizes, some less than a half inch in diameter. HF tags can be easily read while attached to objects containing water, tissues, metal, wood, and liquids. They are affected by metal objects in the close vicinity, however. HF interrogators are less complex than UHF interrogators and usually cost less than the latter.

Inductive coupling used by the HF interrogators uses magnetic flux to power and communicate with the tags. Magnetic flux is omnidirectional (not directional) and therefore it covers the entire area, surrounding its source evenly. That means there are no holes in its density. This makes HF tags an ideal choice for applications such as a smart shelf, for which the entire shelf area needs to be covered with magnetic flux so all the items on the shelf can be interrogated. Other applications of the HF tags include credit cards, smart cards, library books, airline baggage, and asset tracking. Due to the absence of restrictions on the use of the HF frequency and the popularity of smart cards, HF tags are currently the most widely used tags around the world. Figure 3-10 shows HF tags.

FIGURE 3.10 HF tags

Exam Tip

You should know that an HF tag has an antenna with three to seven turns of coil, while an LF tag antenna has several hundred turns of coil.

Ultra High Frequency (UHF) Tags

The ultra high frequency range includes frequencies from 300 to 1000 MHz, but only two frequency ranges, 433 MHz and 860–960 MHz, are used for RFID applications. The 433 MHz frequency is used for active tags, while the 860–960 MHz range is used mostly for passive tags and some semi-passive tags. The frequency range of 860–960 MHz is often referred to by a single frequency of 900 or 915 MHz. Tags and interrogators in this range are called UHF tags and UHF interrogators. The passive and the semi-passive tags in this frequency range use far-field radiative coupling, or backscatter coupling. The UHF tags have a read range of about 15 to 20 feet. All the protocols in the UHF range have some type of anti-collision capability, allowing multiple tags to be read simultaneously in the IZ. The new Gen 2 protocol for UHF tags is designed for reading several hundred tags per second. UHF interrogators are usually costlier than HF interrogators, but UHF tags are becoming more economical. Figure 3-11 shows some UHF tags.

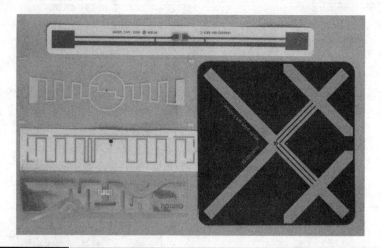

FIGURE 3.11   UHF tags

The UHF tag antennas are usually made of a copper, aluminum, or silver deposited on the substrate. Their effective length is approximately 6.5 inches, which is approximately equal to one-half the wavelength of 900 MHz radio waves. The optimum length of an UHF antenna is equal to one-half the wavelength of the carrier wave, though with proper design, the length can be reduced. The UHF antennas are thin and easy to manufacture, allowing tags to be very thin, less than 100 μM, almost two-dimensional. The UHF tags cannot be easily read while attached to objects containing water and animal tissues because water absorbs UHF waves as well as detunes the tag. The UHF tags also get detuned when they are attached to metal objects. Separating UHF tags from the metal objects or objects with liquid improves their performance. UHF tags cannot be read if water or any conductive material is placed between the interrogator antenna and the tags.

Radiative coupling used by the UHF interrogators uses radio waves to power and communicate with the tags. Reflection, diffraction, and refraction of radio waves gives rise to a multipath effect, where radio signals arrive at the receiver using multiple paths. Some of the signals from multipath attenuate the original signal. This creates an IZ with varying signal strength. The tags may not be readable in low signal spots, causing random tag readability problems. The UHF antennas are directional, which helps create an IZ with well-defined boundaries, though the zone may have holes in it. UHF tags are getting a big boost from the mandates by large organizations for their use in the supply chain. This and the creation of the Gen 2 protocol have created a tremendous momentum in RFID industry to manufacture low-cost UHF tags in high volume.

UHF frequency regulations are not as uniform as HF frequency regulations. Governments in various parts of the world had assigned the UHF frequencies at around 900 MHz, long before RFID, for uses other than for RFID. Therefore, there is no common frequency range around 900 MHz available for RFID use. Different countries have different bands available with varying allowable maximum power levels and duty cycles. To overcome this problem, the Gen 2 protocol was designed to work with any frequency band within the 860–960 MHz range and with different maximum power levels. The government regulations divide the allocated frequency range into a number of narrower frequency bands. These narrower bands are called *channels*. Different countries have different numbers of channels available within their allocated bandwidths. The regulations also require that the interrogators do not use a single channel all the time but pseudorandomly hop among the available channels.

The following table shows allocated band size, maximum allowable power, and number of channels allocated in a few countries. Some countries use effective isotropic radiated power (EIRP) for maximum power specifications, while the others use effective radiated power (ERP). The difference between them is discussed in Chapter 2.

Region	Band Size (MHz)	Maximum Power	No. of Channels
North America	902–928	4 W EIRP	50
Europe (302–208)	865–868	2 W ERP	15
Japan	950–956	4 W EIRP	12
Singapore	866–869	0.5 W ERP	10
	923–925	2 W ERP	
Korea	908.5–914	2 W ERP	20
Australia	918–928	4 W EIRP	16
Argentina, Brazil, Peru	902–928	4 W EIRP	50
New Zealand	864–929	0.5–4 W EIRP	Varied

Microwave Tags

The microwave frequency range includes frequencies from 1 to 10 GHz, but only two frequency ranges around 2.45 GHz and 5.8 GHz are used for RFID applications. Almost all microwave tags use 2.45 GHz. Microwave tags are available as passive, semi-passive, and active types. The passive and semi-passive tags use backscatter coupling to communicate with interrogators, and active types use their own transmitter to communicate. Passive microwave tags are usually smaller than passive UHF tags and have the same read range of about 15 feet. The semi-passive microwave tags have a read range of about 100 feet, while the active microwave tags have read range of about 350 feet. Passive microwave tags, due to low demand, are more expensive than passive UHF tags, but they share the same advantages and disadvantages. Only a few manufacturers make this type of tag. Japan is the largest user of passive microwave tags.

Microwave antennas are directional, which helps define the IZ for passive and semi-passive tags. Due to their shorter wavelength, they are easier to design to work with metallic objects. A wider band of frequencies is available to use and more hop channels are available. However, many commonly used devices such as cordless phones and microwave ovens use this frequency. Therefore, interference at microwave frequencies is possible. Government regulations regarding use of microwave frequencies for RFID are almost nonexistent. The semi-passive microwave RFID tags are used in long-range access control for vehicles, fleet identification, and highway toll collection. Active microwave tags are used for real time location systems (RTLS). Figure 3-12 shows a microwave tag.

Exam Tip	
Remember that UHF and microwave tags have dipole antennas.	

Tag Communication Methods

The interrogators and tags communicate using various methods, depending on tag design parameters such as frequency of operation and source of power. Passive and semi-passive UHF and microwave tags use radiative coupling, and passive LF and HF tags use inductive coupling for communication. Active UHF and microwave tags include their own transmitters. The communication method affects the performance of the tag, dictating the maximum read range and consistency of read within that range.

FIGURE 3.12   Microwave tag

The inductive coupling employed by LF and HF tags uses magnetic flux to power the passive tags and communicate with them. The electromagnetic flux strength of inductive coupling decreases by the sixth power of the distance between the tag and the antenna. This provides a very short read range for inductively coupled tags. The radiative coupling employed by UHF and microwave tags uses radio waves to power the passive tags and to communicate with them. Radiative coupling provides a longer read range than inductive coupling because the radio wave strength decreases by only the second power of the distance between the antenna and the tag. Radiatively coupled tags use modulated backscatter on a downlink communication. (Backscatter was discussed in Chapter 2.) Active tags use their radio at the designated frequency, usually 433 MHz or 2.4 GHz, to transmit data with or without the presence of an interrogator. They can transmit and receive data to and from other active tags within their radio range.

Functionality: EPCglobal Tag Classes

Standards organization EPCglobal has categorized RFID tags into six different classes according to the following functionalities implemented on the tag:

- Write capability
- Power source
- Memory capacity
- Communication capability

These tag classes are numbered from 0 to 5. The following table provides a brief description of each class:

Class	Description
Class 0	Passive, data written once during manufacture, read only
Class 1	Passive, factory or field programmable once only, read only thereafter
Class 2	Passive, read/write, user memory and encryption
Class 3	Semi-passive, on-board sensors, read/write, user memory
Class 4	Active, read/write, on-board sensors, peer-to-peer communication with other active tags in same frequency band and with interrogators
Class 5	Essentially the interrogator, read/write, can power classes 0, 1, 2, and 3 tags, can wirelessly communicate with all the classes

The classes 0, 1, and 2 are all passive tags with various write and memory capabilities. The class 3 tags are semi-passive tags and may have an on-board sensor. All the tags in classes 0, 1, 2, and 3 can communicate only with class 5. The class 4 tags are active tags and can communicate with any class 4 or 5 device. A class 5 device is an interrogator and can communicate with any class.

Over time, some manufacturers have added functionalities to tags and started assigning their own class numbers, such as class 0+ tags, making the distinction between passive tag classes 0 through 2 become somewhat fuzzy. As you will see in the next section, even the EPCglobal Gen 2 protocol does not follow the five-class classification: semi-passive tags are still referred to as class 3 tags and active tags are referred to as class 4 tags.

Protocols

Protocols are the definition and grammar of the language used by the interrogators and tags to communicate with each other. The interrogator and tag must use the same protocol to communicate. The interrogator may have multi-protocol capability, but it uses only one protocol at a time. Typically, a multi-protocol interrogator is set to a single protocol, or it may cycle through several protocols one by one. Protocols may be designed by a manufacturer or by a standards organization. Manufacturer designed protocols are usually proprietary and may not be available to all the manufacturers. Protocols designed by a standards organization are called *open protocols* and are available on equal terms to all manufacturers. For this reason, it is always a good idea to purchase equipment that uses an open and popular protocol; it increases the availability and compatibility of the equipment and helps protect your investment. The protocols define *air interface*—that is, how a tag and an interrogator communicate using electromagnetic waves. This includes frequency of operation, emission power level, data rate, signal modulation, encoding of data bits, data structure, command structure, and anti-collision algorithm.

Examples of proprietary protocols include Philips I-Code, TI Tag-It, Alien EPC Class 1, Matrics EPC Class 0, and Intermec IntelliTag. Examples of open protocols are ISO 14443 (A/B), ISO 15693, ISO 18000-6 (A/B), and EPCglobal Class 1 Generation 2 (Gen 2).

Gen 2 Protocol

The official name of this protocol is Class-1 Generation-2 UHF RFID Protocol, but it is usually referred to as Gen 2, C1G2, or Generation 2 protocol. It was developed by EPCglobal, Inc., and published in December 2004. It was submitted to the International Organization for Standardization (ISO) in March 2005 for approval as an ISO standard and was approved as ISO 18000-6C standard in July 2006.

More than 50 companies from various industries provided input for development of Gen 2 protocol. It is designed to work globally regardless of the local UHF regulations. This is the most important aspect of this protocol, because UHF regulations vary from country to country. In global commerce, items move from country to country; therefore, the tags attached to the items in one country must be readable in any other country regardless of that country's regulations. The UHF range was broadened to include regulations of all the countries. The Gen 2 protocol is designed to work at frequencies from 860 to 960 MHz with various levels of power and duty cycle. Gen 2–compliant interrogators are designed to work in a particular country according to that country's UHF radio regulations. Gen 2–compliant tags are designed to work at any frequency between 860 and 960 MHz. Therefore, Gen 2 tags can be read by any Gen 2 interrogator in any country.

The Class 1 Generation 2 protocol defines tags to have multiple read/write capability. According to EPC tag classification, multiple read/write capability belongs to Class 2. This shows that with the creation of the Gen 2 protocol, the passive UHF tag classifications have taken a back seat. Gen 2 is an open protocol that promises interoperability between interrogators and tags designed and manufactured by different manufacturers anywhere in the world. With the availability of open and globally applicable protocols, all today's manufacturers are producing Gen 2 interrogators and tags.

EPCglobal has created a Gen 2 compliance certification program whereby manufacturers can get their products certified for Gen 2 compliance. A Gen 2–certified interrogator from one manufacturer will be able to read Gen 2–certified tags from another manufacturer. The Gen 2 protocol is discussed in more detail in Chapter 10.

Objective 3.3 **Tag Performance**

T ag performance is measured using many different parameters, including read rate, read distance, yield, and consistency. All these parameters are affected by tag design, tag manufacturing processes used and quality control during manufacturing, tag orientation, tag placement, material tagged, tag motion in the IZ, and environmental characteristics of the IZ. With so many variables affecting tag performance, it is difficult to evaluate tag performance. It requires expertise in electromagnetic wave communication and a proper test design and setup. The following sections discuss the effects of various parameters on tag performance.

With the current manufacturing processes, tag quality can vary from one batch to the next or even from one tag to the next. This makes it difficult to extrapolate results of tag tests to a larger population of identical tag types. The best way to deal with this situation is to test several tags and average of their performance. This, of course, makes tag testing a tedious and resource-intensive activity.

Power Source

RFID tags may be powered by an on-board battery or by an external power source. The external power is usually delivered via magnetic flux or radio waves. The magnetic flux creates a uniform density zone, allowing tags to be powered consistently within the zone. Radio waves, used by UHF and microwave tag, create a zone of radio signals that is affected by reflected waves from surrounding objects. The field density varies unpredictably within the zone. Therefore, depending on the tag location within the zone, the tag may or may not receive strong enough radio waves to power itself. The active tag with an on-board battery transmits radio waves with a consistent strength, except when the battery strength decreases. The battery power may decrease because of normal power drain or due to low temperatures. In subfreezing temperatures, battery efficiency greatly decreases, which may cause the active tags to transmit signals with such a low strength that the interrogator may not receive a strong enough signal to extract data, making the tag unreadable.

Tag Orientation and Location

Tag performance is affected by the orientation of the tag relative to the interrogator antenna. The best tag orientation occurs when the tag plane and the antenna plane are parallel to each other. At this orientation, the tags receive the maximum power. As the tag is rotated, it presents a smaller effective area to the incoming radio waves and thus collects less power. The tag read range decreases as the collected power decreases.

Most passive UHF tags have a single dipole antenna that has nulls along its north and south poles. This type of tag cannot be read when the axis passing through the poles of the antenna is perpendicular to the plane of the interrogator antenna. Two options are available to remedy this situation: install two interrogator antennas—one perpendicular to the other—so the tag antenna axis is not perpendicular to the plane of at least one of the interrogator antennas. Or, use tags with two dipole antennas (a *dual dipole tag*), with one antenna perpendicular to the other. The axis of at least one of the two antennas on this tag will never be perpendicular to the interrogator antenna plane.

The location of the tag within the IZ also affects the tag's performance. As the tag moves away from the interrogator antenna, it receives less and less power. The reflected signals mixing with the original signal further reduces the power available to the tag. At the outer edges of the IZ, signal strength may decrease in pockets to such a low degree that the tag may not be powered and the consistency of the tag read decreases.

Tag Placement

Placement of the tag on an object affects the tag's performance. Placement is an important consideration for passive UHF and microwave tags used for the products containing aqueous liquids, such as wine bottles, shampoo bottles, many drugs in liquid form, and most food items. Radio waves at UHF and microwave frequencies are absorbed by the water. To read UHF tags attached to packaging containing aqueous liquids, the tag must be placed so that an air gap exists between the tag and the aqueous liquid. You can design the tag so that its antenna is always pointed away from the liquids, or you can take advantage of the way the liquids are packaged so that the tag is attached to the container away from the liquids in the container. At case-level tagging, as opposed to item-level tagging, the tag is attached to the case. The best place to attach a passive UHF tag on cases is where the item packaging provides the most separation from the liquids inside. This may require a very precise tag placement, with a tolerance of only a few millimeters. The best way to determine the best tag location is to experiment with various tags and locations. You can also use services of RFID labs, which can help you determine the best tag placement on your product using various test methods.

Tag Stacking

Tag stacking occurs when several tags are placed close to each other in almost an identical orientation, such as when small items are tagged and then packed in a case. In this situation, the tags in the front of the case, closer to the antenna, absorb and reflect most of the radio energy, so the tags on the middle items receive a very weak signal and are not powered up, and thus cannot be read. To avoid this problem, you can change the packaging method or rotate the case in front of several antennas installed at various angles to the case. Tag stacking is also referred to as tag shadowing.

Antenna Polarization

Interrogator antennas are designed to be linearly or circularly polarized. Linearly polarized antennas can be installed in a horizontal or vertical polarization position. Circularly polarized antennas can have right-hand or left-hand circular polarization. For a linearly polarized antenna, orientation of its polarization must match with the orientation of the tag. A tag in a horizontal position therefore cannot be read by a vertically polarized antenna and vice versa. The circularly polarized antenna can read a tag in any orientation but at the expense of the amount of power received by the tag. A circularly polarized antenna splits transmitted power into two planes, one perpendicular to the other. A tag therefore receives only half the power relative to what it would receive from a similar linearly polarized antenna, which in turn reduces the tag read range. You gain tag orientation insensitivity but you lose some read range.

Circularly polarized antennas are most common in dock door applications, where tag orientation cannot be controlled. Linearly polarized antennas are sometimes used for conveyors in manufacturing environments. If the read range is not a factor, it is best to use circularly polarized antennas when possible.

Tag Speed

The speed at which tags pass through the IZ affects the number of tags that can be read in a particular time interval (usually measured in seconds). The length of time a tag spends in the IZ is called *dwell time*. By knowing the dimensions of the IZ and direction and speed of tag travel, you can calculate dwell time. The interrogator specifications should supply the number of tags it can read in 1 second. This number is usually not very accurate to use however, because the manufacturer does not provide details regarding how that number was calculated. A single tag within an IZ may be read, for example, 1000 times per second, but when multiple tags are placed in the IZ at the same time, that number does not scale. With more than one tag in the zone, the interrogator has to use an anti-collision algorithm to singulate the tags before reading them. The time required by an anti-collision algorithm increases with the increase in the number of tags in the zone. Therefore, for example, a single tag in the IZ may be read 1000 times per second, but 200 tags in the zone cannot be read in 0.2 second; it will take much longer to read this number of tags. The best way to determine how many tags an interrogator can read is by conducting a well-defined and controlled experiment in the operating environment. Once you have a reasonable estimate of this number and you know the dwell time and the maximum number of tags in the zone, you can determine whether you will be able to read all the required tags.

Exam Tip
Know that dwell time = interrogation zone length / travel speed of tag.

Environmental Factors

Environmental factors such as radio interference and humidity can affect tag performance. Radio interference may be constant or random and may come from external or internal sources. The external interference may be due to geography of the facility and use of various radio devices near the facility. These interferences are predictable but not easily controllable. You may have to install radio energy shielding material around the IZ. The internal interference may be due to improper installation or operation of some radio devices, or it may be due to reflection from nearby objects. A proper installation and some fine-tuning at the IZ can help eliminate or reduce these internal interferences. The most difficult to deal with are random occurrences, which require a careful study and may involve much trial and error before they are eliminated.

The humidity in the air may not directly affect the performance of the tag. However, when tags are attached to objects that absorb humidity from air, the tags' performance may vary with humidity. Cold tags passing through high humidity areas may develop condensation, which may reduce tag performance.

Read vs. Write

All tags require more power, sometimes twice as much, to write data than to read it. This reduces the tag's write range. If you have configured the IZ using a maximum read range, you will not be able to write to the tags at this distance. You must design the IZ for writing to the tag differently than your design for reading the tags. Interrogators also require more time to write data to a tag than to read data from a tag. An interrogator may be able to read 1000 tags per second but may write only 10 tags per second. In addition, you may be applying tags or labels while writing to tags. The applicator may apply only one or two tags per second. Therefore, a tag writing station must be designed specifically for that purpose.

Objective 3.4 Tag Selection

Selecting a proper tag is no simple task. It requires a thorough understanding of how tags and interrogators work and an understanding of the business processes performed on the objects to be tagged. You must understand tag and interrogator specifications and their impact on tag performance. Some of the factors considered during tag selection are tag type, operating frequency, materials to be tagged, tag mounting method, read range, read rate, tag size, environmental conditions, cost, and mandated requirements.

Frequencies

Selection of tag frequency depends on factors such as read range requirements, material to be tagged, and data rate. Applications such as access control and payment systems require a very small read range. In these types of applications, reading tags beyond a few inches creates a security risk, and therefore the read range must not extend beyond a few inches. LF and HF tags with shorter read ranges are ideally suited for such applications. In tagging pallets used in a supply chain, a read range of 5 to 10 feet is required. UHF and microwave tags are well suited for this application.

Items with aqueous content require separation between the item and UHF or microwave tag. At the item level, few choices for tag placement are available and less opportunity exists to create an air gap between the tag and the object. In this type of situation, an HF tag with 2 to 3 feet of read range is an ideal choice. HF and LF tags are well suited for tracking humans and animals because living bodies have a high water content. The FDA has approved LF tags for human and animal implantation.

Objects with metals pose a special problem to tagging. Tags directly attached to metal objects are detuned and cannot work well regardless of the tag frequency. If UHF and microwave tags are attached to metal with a small air gap between the tag and the object, they can be easily read. LF and HF tags are more sensitive to metal and require a larger air gap.

Tag Type

The factors affecting tag type selection are read range requirements, sensor requirement, cost, size, weight, and the type of application. Passive tags are cheaper than semi-passive or active tags. Tag cost should be considered relative to the cost of the items being tagged. For example, a $100 active tag attached to a container is economical, while a 25 cent tag attached to a box of cereal may not be. Some applications dictate the type of tags used. For example, to monitor temperature within a refrigerated truck, you would need semi-passive or active tags with temperature sensors. To monitor security and integrity of a shipping container, an active tag with an intrusion sensor may be used. A reusable tote or plastic container used in a manufacturing or food-processing plant may be fitted with a semi-passive tag. Since these totes and containers re-circulate inside the plant, the cost of the tag is not a major concern.

> **Travel Advisory**
> When selecting a tag, consider the cost of applying the tag to the object.

Environmental Factors

The environmental conditions that the tag may encounter during its lifetime are major considerations in tag selection. A tag embedded within the product may encounter high temperatures and pressures during product manufacturing, and it must be able to survive those conditions. Evaluate environmental conditions not only within your own facility but also in any environment the tag will travel through during its entire lifetime. These environmental conditions may affect the tag's read range or read rate:

- The substrate may absorb moisture or may become brittle and crack.
- The adhesive may not hold the tag due to moisture or chemicals and the tag may fall off the item.
- The interconnection between the antenna and the IC may break due to vibration or due to repeated flexing of the tag.
- The antenna may be weakened due to corrosive effects of the chemicals in the environment.

You may need to encapsulate tags to protect them from abrasion or chemicals in the environment.

RFID Standards Compliance

Tags may be selected for compliance with certain standards. The standards may be open, such as ISO, or they may be proprietary. It is a good practice to follow the open and popular standards, because more companies manufacturing products use these standards than proprietary ones, providing a wider choice of products. In addition, these products may be cheaper due to higher volume, and support for them may be available for a longer period. With the wide acceptance of Gen 2 standards, it is best to use Gen 2–compliant products if you are going to use UHF.

Compliance Requirements

Many large organizations and government agencies have mandated that their suppliers provide goods with RFID tags. These published mandates may specify tag type, frequency, amount of memory, read range, read rate and speed, and protocol. In addition, the mandates may specify how the goods should be tagged. If you are supplying goods to one of the mandating companies, you have little choice in tag selection. The frequency and tag types are specified in the mandates. You may select tag manufacturer and tag model depending on the materials being tagged and the location of the tag on the goods.

Typical Tag Specifications

This section looks at some typical tag specifications and discusses their implication to tag selection.

The following table shows specifications for a passive UHF tag from Texas Instruments:

IC Standards	EPC UHF Gen 2
Operating frequency	860–960 MHz
EPC memory	96 bits user programmable
Tag ID memory	32 bits factory pre-programmed
Write/erase cycles	1000 at 25° C
Operating temperature	–40° C to 65° C
Storage temperature (single)	–40° C to 85° C
Storage temperature (on reel)	–40° C to 45° C
Bending radius	0.59 in. (15 mm)

Antenna size	3.5 in. × 1 in.
Inlay pitch	1.5 in.
Inlay width	3.75 in.
Die height	~11 mils
Substrate	75 micron (~2.95 mils), PET
Antenna material	Printed silver ink
Reel diameter	ID: 3 in. core; OD: max 15 in.
Delivery	Single row inlay wound on cardboard reel
Quantity	10K per reel

These specifications provide details about the physical size of the tag. The UHF tags are EPC Gen 2 compliant and have 96 bits of memory in which the user can write data. Data can be written to the tag 1000 times. Although some manufacturers quote 10,000 times, a typical UHF tag data may need to be written to only a few times, so 1000 times is more than adequate for the vast majority of applications. The storage temperature of tags on a reel is lower than the storage temperature of a single tag because the expansion due to higher temperatures may stress the tags on the reel. Minimum bending radius is provided for use of a tag roll in high-speed equipment. Die height is the height of the IC above the substrate.

The following table shows specifications for a semi-passive tag from Alien Technologies:

Frequency	2450 MHz (ISM band)
Bit rate	16K bits/second
Range	Up to 100 feet
Memory	4K non-volatile read/write
I/O	Real-time clock and on-board temperature sensor Range –55° C to 125° C Accuracy +/– 2.0° C
Power source	3 volt 220mA/h lithium battery
Operating temperature	–25° C to +70° C
Storage temperature	–40° C to +85° C
Mechanical shock	IEC 68-2-27
Vibration	IEC 68-2-6
Dimensions	3.15 × 1 × 0.6 inches

This tag operates at 2.4 GHz and can transfer 16K bits of data in 1 second. Its maximum read range is 100 feet. It has 4K of memory that can retain data even if the battery on the tag is dead. It has an on-board temperature sensor with a good low temperature range, so this tag may be a good choice for monitoring frozen goods. For shock and vibration, it conforms to International Electrotechnical Commission (IEC) standards.

The following table shows specifications for an active tag from Savi Technologies:

Dimensions	6.25 × 2.125 × 1.125 in. (15.875 × 5.4 × 2.86 cm)
Weight	3.8 ounces (108 grams)
Case material	Federal Standard 595 #34094-Bronze Green
Beeper	Audible beeper for tag location
Mounting	Rivets, Pressure Sensitive Tape, Pre-installed on Mounting Sleeve
Temperature	–32°C to +70°C Operating; –40°C to +70°C Storage
Humidity	100% Noncondensing
Vibration and shock	MIL-Std-810E Method 15.4, Category 10
Enclosure	Sealed to IP 54
Frequency	433.92 MHz (UHF Transceiver)
Range	Up to 300 feet, unobstructed when mounted on container
Data rate	27.8
Transmit power	0.6 mW
Frequency	123 kHz (LF Receiver)
Range	Up to 12 ft. (3.65 m) from Savi Long Range Signpost
Battery type	3.6 V lithium, replaceable by user without tools
Battery life	Approximately 5 years at two collections per day
Memory	Available with 128K of user memory
Type	Unlicensed operation under FCC Part 15 ETSI, EN 300 220-1 EN 300 330

This tag operates at two frequencies. At 433.92 MHz, it transmits and receives data and has a range of up to 300 feet, while at 123 kHz it only receives data and has a range of 12 feet. It has a beeper to locate the tag. Many specifications are provided as conforming to applicable standards. It has a replaceable battery with a five-year life if the data is collected only twice a day. A more frequent data collection will provide a shorter battery life. It has 128K of user programmable memory. This is a lot of memory (compare that to the original IBM PC, with 64K of memory).

Travel Advisory

Never select a tag based only on its specifications. Always test several samples with the objects to be tagged in the actual environment in your facility.

CHECKPOINT

✔ **Objective 3.1: Tag Components and Construction**   RFID tags (also called inlays) are made of three different components: an integrated circuit (IC), an antenna (coil or dipole antenna), and a substrate. An IC can be attached to an antenna either directly or using a strap. Inlay can be used "as-is" as an insert, attached to a label with adhesive to create a smart label, or embedded in RF-transparent material to form an encapsulated tag.

✔ **Objective 3.2: Tag Types**   Tags come in three different types: passive, semi-passive, and active. Passive tags do not use a battery or transmitter. They collect power from the reader. Semi-passive tags have an on-board battery to power them but have no transmitter. In addition, they may contain an on-board sensor. They have a read range of up to 100 feet. Active tags have an on-board battery and a transmitter. Their read range is up to 600 feet. They may transmit data with or without the presence of an interrogator.

Tags also vary by operating frequency and communication methods. Inductively coupled tags operate at LH and HF. Their antennas are in the form of a coil with three to several hundred turns. Their read range is less than 3 feet. Backscatter coupled tags operate at UHF and microwave frequencies. Their antennas are usually dipole type and their read range is up to 15 feet. Inductively coupled tags use magnetic flux while backscatter coupled tags use radio waves.

Tags can be classed according to standards and protocols. Tags can follow either open (such as ISO or EPC) or proprietary (manufacturer) standards. EPC tags fall into classes according to their capabilities; the newest is the EPC Class 1 Generation 2.

✔**Objective 3.3: Tag Performance**   The orientation of the tag relative to the interrogator antenna affects the tag's read performance. The best tag orientation is a tag plane parallel to the antenna plane. The worst orientation is a tag antenna axis that is perpendicular to the antenna plane. The location of a tag within the IZ dictates how far the tag can be read, because the strength of radio waves decreases as the antenna is moved away from the axis perpendicular to the antenna plane. Circularly polarized antennas are less sensitive to tag orientation. The tag orientation must match with a linearly polarized antenna for the tag to be read.

The content of the package to be tagged affects the read performance of the tag. If the package contains objects with aqueous substances, the substance will absorb the radio waves and decrease tag read range. Metal in the package may detune the tag and reduce the tag performance. Both metal and aqueous substances in the package will not allow radio waves to pass through. This would affect other tags behind such a package.

✔**Objective 3.4: Tag Selection**   Many tag selection criteria include tag type, operating frequency, materials to be tagged, tag mounting method, read range, read rate, tag size, environmental conditions, cost, and mandate requirements. It is a recommended practice to test the tags on the objects to be tagged. Tag selection involves compromise among conflicting criteria.

The substrate and the adhesive on the tag affect the quality and performance of the tag. The substrate may absorb moisture or chemicals and reduce read performance of the tag or may corrode the tag antenna. The substrate may dry up and crack. The adhesive must be selected so it holds the tag to the object under different environmental conditions. It must also be compatible with the tag use—for example, the adhesive of the tag placed on food items should not be toxic.

REVIEW QUESTIONS

1. What is the purpose of the integrated circuit (IC) on the tag?

   A. It collects radio energy.

   B. It provides a place to store data.

   C. It isolates a tag from its surroundings.

   D. It reflects radio waves.

2. What does a low frequency (LF) tag antenna do?

   A. It backscatters the radio wave.

   B. It couples with radio signals from the antenna.

   C. It couples with the interrogator antenna using magnetic flux.

   D. It transmits radio waves.

3. Which of the following materials can be used for a tag substrate?

   A. 75 micron thick aluminum sheet

   B. 25 micron thick aluminum sheet

   C. 75 micron thick PET sheet

   D. 25 micron thick ceramic sheet

4. In a supply chain, smart labels are typically used to tag what?

   A. Cases of items

   B. Cans of soda

   C. Milk cartons

   D. Cereal boxes

5. Which of the following is not a part of a semi-passive tag?

   A. Antenna

   B. Battery

   C. IC

   D. Transmitter

6. Which of the following tag types use backscatter?

   A. Passive and active tags

   B. Passive and semi-passive tags

   C. Battery assisted tags and active tags

   D. Semi-passive and active tags

7. Which of the following frequency ranges use inductive coupling?

   A. LF and UHF

B. LF and HF

C. UHF and microwave

D. UHF and HF

8. Which of the following frequency range is best suited for reading objects with high aqueous contents from a distance of approximately 2 feet?

   A. HF

   B. LF

   C. UHF

   D. Microwave

9. The Gen 2 specification was written for which of the following frequency ranges?

   A. 860–960 kHz

   B. 902–928 kHz

   C. 902–928 MHz

   D. 860–960 MHz

10. For an application requiring 75 feet of read range, which of the following is the best type tag to use?

    A. Passive UHF tag

    B. Passive HF tag

    C. Semi-passive HF tag

    D. Semi-passive UHF tag

11. The active tags belong to which of the following classes?

    A. Class 4

    B. Class 0+

    C. Class 3

    D. Class 5

12. Three ISO 18000-6A tags, two ISO 1800-6C tags, ten HF tags, and four LF tags are used in an IZ with one UHF Gen 2 interrogator and four antennas installed. What is the total number of tags the interrogator will read?

    A. 5

    B. 14

    C. 2

    D. 3

13. Which of the following is the best tag orientation for reading UHF tags?

   A. The antenna plane and tag plane are parallel to each other.

   B. The antenna plane and tag plane are perpendicular to each other.

   C. The antenna plane and tag plane are at 90 degrees to each other.

   D. The tag axis is perpendicular to the antenna plane.

14. An UHF interrogator is connected to a horizontally polarized antenna with 6 dBi gain. This antenna is replaced by a 6 dBi gain circularly polarized antenna. Which of the following best describes the consequences of this change?

   A. The tag in the horizontal orientation cannot be read.

   B. The tag in the vertical orientation cannot be read.

   C. All the tags can be read.

   D. The tags that were near the borderline of the IZ cannot be read.

15. Tag dwell time is defined how?

   A. The time to read one tag

   B. The time to read all the tags in the IZ

   C. The time the tag stays inside the IZ

   D. The time between two successive interrogation attempts

REVIEW ANSWERS

1. **B** The integrated circuit (IC) has memory to store data. The object ID is stored in this memory.

2. **C** A LF tag antenna couples with the interrogator antenna using magnetic flux.

3. **C** A 75 micron thick PET sheet can be used for tag substrate. The ceramic sheet will be too fragile to use.

4. **A** Cases of items are usually tagged with smart labels.

5. **D** Semi-passive tags do not have a transmitter.

6. **B** Passive and semi-passive tags use backscatter to communicate with the interrogator.

7. **B** LF and HF ranges use inductive coupling.

8. **A** HF is the best frequency range to use for objects with aqueous substances and at a distance of approximately 2 feet. The LF could read

objects with aqueous substances but will not provide a read range of 2 feet.

9. **D** The Gen 2 specification covers the frequency range 860–960 MHz.

10. **D** Semi-passive UHF tags provide a read range of up to 100 feet.

11. **A** Active tags belong to EPC Class 4.

12. **C** The interrogator will read only two tags because it is designed for use with the Gen 2 protocol. Only two tags use the ISO 18000-6C protocol, another name for Gen 2. A tag and interrogator must use the same protocol to communicate.

13. **A** When the antenna plane and tag plane are parallel to each other, the tag is in the best orientation for reading as well as for writing. It collects the maximum power.

14. **D** The tags that were near the borderline of the IZ cannot be read because the circularly polarized antenna transmits only half the power of a linearly polarized antenna. This will reduce the size of IZ to about 70 percent and the tags that fall beyond this cannot be read.

15. **C** Dwell time is defined as the time the tag stays inside the IZ.

Interrogation Zone

ETA	NEWBIE	SOME EXPERIENCE	EXPERT
	6 hours	3 hours	1 hour

An interrogation zone (IZ, shown in Figure 4-1) consists of the interrogator, antennas, cables, peripherals, and the environment in which the equipment is installed. The complexity of an IZ is determined by its environment, in which the presence of multiple objects can cause reflections, absorption, or interference with the original signal. These factors can cause unintentional reads, can block the intended reads, and can reduce throughput processing capability.

Hardware and software for an IZ must be properly selected and configured to eliminate or overcome the environmental and business procedural obstacles and limitations. Before you can create a properly functioning IZ, you must understand its components. This chapter starts by explaining the interrogator functionality and its components, and then it reviews various types of interrogators. Next, interrogator antennas, types, polarization, and function are discussed. The next part of the chapter is dedicated to interrogator operation, including air interface and data protocols, dense interrogator mode, anti-colli-

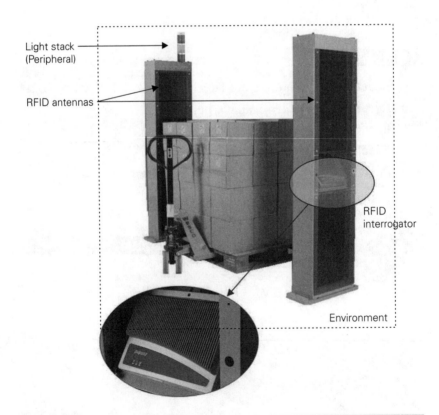

Light stack (Peripheral)

RFID antennas

RFID interrogator

Environment

FIGURE 4.1   Parts of an interrogation zone

sion, and other tag management practices. The last part of the chapter discusses IZ configuration as well as interference problems that could arise when optimizing a system.

Interrogator Functionality

Objective 4.1

Interrogators read and write data to and from tags. Interrogators are responsible for remotely powering the tags (in passive and semi-passive systems), establishing the bidirectional data flow between themselves and the tags, and performing analog to digital as well as digital to analog signal conversion.

Local Lingo

Transceiver Interrogators are commonly referred to as *readers* or sometimes *transceivers.*

Interrogators can also run attached to peripheral devices, such as light stacks or alerting horns that provide feedback on conditions within the system and triggering devices, which control when the interrogators are turned on and off. Peripherals are usually attached to the I/O ports of the interrogators.

Interrogators can also communicate with and control nearby sensors integrated within the IZ and are responsible for communications over networks for alerting back-end systems of the assets or goods they are tracking.

According to their capabilities, interrogators are often called either *smart* or *dumb*. Smart interrogators carry built-in computers that run programs, which filter, aggregate, and analyze data to turn it into meaningful events for the back-end system. A dumb interrogator reads tag data and reports what it sees to the back-end system. This type of interrogator is highly dependent on a middleware product that performs filtering and aggregation functions and converts the data read into a meaningful format, which acts as the input to the back-end system.

Interrogators often have a graphical user interface (GUI) by which they can be controlled. An interrogator GUI is typically accessible through a web interface using any commercial Internet browser on the computer or through a simple application program. Following are some typical features of a GUI:

- Read from and write to tag capability
- Network configuration settings
- Antenna settings
- Power settings
- Input/Output port configuration settings
- Firmware upgrade capability

Firmware upgrades on the interrogator are performed to enhance performance, support new features, or resolve issues with the existing platform. They are usually downloaded from the manufacturer's Web site and can be deployed to the interrogators in a variety of methods. Most common methods are a Pull method, where the new firmware is downloaded or "pulled" by the reader using File Transfer Protocol (FTP) or Hypertext Transfer Protocol (HTTP), and an automated Push method, where the new firmware is automatically uploaded or "pushed" to a reader from the host. The method used depends on the manufacturer and middleware used in the installation.

Interrogator Components

An interrogator includes a transmitter, a receiver, and a processor. The transmitter part of an interrogator consists of the following:

- **Base band transmitter**   Sends the carrier wave that powers the tags
- **Oscillator**   Produces the alternating current that makes up the waves transmitted to the tag
- **Power amplifier**   Amplifies the signals produced by the oscillator
- **Modulator**   The original RF wave's amplitude, frequency of transmission, or phase

The receiver part of an interrogator consists of the following:

- **Amplifier**   Amplifies the weak signals received from the tag before demodulation
- **Demodulator**   Compares the modulated signal to the original signal, thereby extracting the information embedded within

A processor embedded within the interrogator provides these functions:

- Controls network communications with the middleware or back-end system
- Runs the primary operation systems for the interrogator
- Controls the functions of the interrogator
- Controls the memory (ROM, RAM, hard drive type data storage)

Today, many of these components are contained within a digital signal processor so that the radio portion of the interrogator is essentially software-controlled and features can be changed or upgraded on the fly.

The external interfaces to the interrogator are various communication ports such as RS-232, RS-245, or RS-422 serial ports; universal serial bus (USB); Bluetooth and 802.11 wireless network interfaces; LAN ports; general-purpose I/O ports; and most importantly antenna ports typically found in batches of four or eight antennas, depending on whether the interrogator is monostatic or bistatic (discussed shortly).

Interrogators can be powered in different ways. For installations of almost any commercially available interrogator, you will need readily accessible 110 V or 240 V AC outlets (for US and European installations, respectively). Handheld interrogators can utilize an internal battery, but they will have to be recharged using a docking station connected to an AC outlet. In certain situations, it may be appropriate to use Power Over Ethernet (POE), which allows the same wire that transfers data to the interrogator to provide power to it also.

Travel Assistance
For more information selecting frequencies, tags, interrogators, antennas, cables, peripherals, and software, see Chapter 6.

Monostatic vs. Bistatic Interrogator

A *monostatic* interrogator uses a single antenna to transmit and/or receive data to or from a tag. A circulator that lives inside the interrogator turns the antenna from the transmitting to receiving function in a very rapid manner. A *bistatic* interrogator uses two antennas, each dedicated either to transmitting or receiving. Bistatic interrogators usually have twice as many antenna ports as monostatic interrogators.

Objective 4.2 Interrogator Types

Interrogators come in multiple formats, and you choose the most appropriate type to match the business process in which you are trying to use this technology. *Fixed interrogators* are designed to be bolted to walls near doorways or attached to wire racks near doors, integrated into stands and dock door portals, and attached to conveyor portals and the like. *Mobile interrogators* are handheld or mounted to vehicles.

Mobile interrogators, such as handheld interrogators, are often used for low-volume reading or writing in exception processing, quality assurance, and mobile shipping receiving units. Handheld interrogators come in a variety of forms and have a vast selection of options for communications. They can be tethered, receiving power and accomplishing data transmission via a nearby base unit using an extensible cord. Other types of handhelds can be wireless, depending on wireless networks or Bluetooth communications for data transfer, limiting operational time to battery life between charges. Newer renditions of handheld interrogators are appearing in forms of cellular phones and personal digital assistants (PDAs).

Handheld interrogators are usually monostatic, with integrated linear antennas. The monostatic design keeps their size to a minimum since only one antenna is used, while the linear antenna polarization is used to get the best read range. The interrogator's orientation can be changed simply by turning your wrist.

Vehicle-mounted interrogators, which can automate shipping and receiving of goods, are usually integrated into material handling devices such as forklifts, paper trucks, cargo trucks, and pallet jacks. These interrogators usually have a special shape for easier installation onto the vehicle and a rugged design to survive the vibrations and other environmental conditions.

Interrogators can also come in the form of a PCMCIA card for laptops, which is capable of being installed into any standard laptop port; or module-based interrogators found in RFID-enabled printers.

> **Travel Advisory**
> Not all laptops include a standard PCMCIA port. Some feature instead the newer ExpressCard.

 **Objective 4.3**   # Interrogator Antennas

The interrogator antenna is a converter between the radiated waves and wired voltage that feeds the system. It is the largest and most obvious and most exposed point of the system, as it must be near the assets it is identifying. This also makes it the most vulnerable point of the system, because it can be physically impacted by goods traversing the IZ. Antennas tend to be near doorways in distribution centers and manufacturing centers so they are occasionally exposed to rain, snow, wind, and high-speed vehicles (such as fork trucks) traveling through the IZ.

Antenna Footprint

The footprint of the antenna will dictate the size of the IZ and how well the zone will be covered by the antenna. Most antennas come with a radiation plot that defines the area the antenna will cover, how far away the signal can be measured, and how wide the antenna beam is at its widest. Overlapping antenna coverage is always a good idea in fixed interrogator installations to ensure consistent and maximum coverage within the IZ.

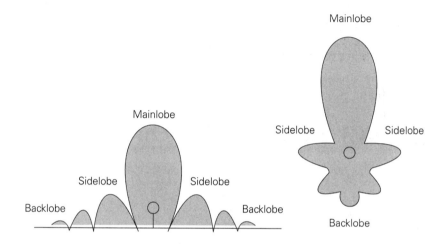

Antenna Polarization

Both linearly and circularly polarized antennas are used in common RFID supply chain applications. *Linear antennas* propagate only on a single plane, so they are either horizontally or vertically polarized. Linear antennas have a more focused beam than a circular antenna, and as a result, more power is realized on a single plane in the IZ compared to circular antennas. Therefore, linear antennas achieve longer read ranges and better penetration of dense objects than circular antennas.

Circularly polarized antennas spread their power equally across a vertical and a horizontal plane; therefore, the signal is not propagated as far and a tag does not receive as much power as it would from a linear antenna, because the power is split across two planes. The main advantage of a circular antenna is, however, that the objects do not need to be in a specific orientation in order to be read.

In a mixed environment where orientation cannot be controlled, circular antennas work best. In a manufacturing environment where orientation can be assured, linear antennas are better because they can penetrate dense objects.

> **Travel Advisory**
>
> Antennas are neither monostatic nor bistatic, as this is determined by the interrogator. However, they are sometimes referred to in these ways because an antenna that is intended for use in a bistatic interrogator usually consists of two antennas in its casing and is larger in size. Theoretically, this antenna could be used as two monostatic antennas if it were connected to two antenna ports on a monostatic interrogator.

 **Objective 4.4** **Interrogator Operation**

Interrogators receive commands from the back-end system to perform various functions, such as reporting what they see either in real time or at fixed intervals throughout the read cycle. Commands to start reading or writing tags at a particular time can also come from triggering devices in the IZ. As mentioned, interrogators transmit power and data to the tags, receive the signal back from the tags, decode the signal, and hand that signal off in digital format for the back-end system. Although this may sound relatively simple, many functions are involved in this process, such as air interface protocols, multi-interrogator environment modes, and anti-collision functions.

Protocols

An *air interface protocol* is the means by which the tag communicates with the interrogator and the interrogator communicates with the tag. *Data protocols* determine how the interrogator interprets the numbers it reads during this exchange. Air interface protocols and data protocols are governed by various standard bodies such as the International Organization for Standardization (ISO) and EPCglobal, the two primary bodies that define how these protocols work.

Proprietary air interface protocols also exist; these are specific to certain industries such as the Automotive Industry Action Group (AIAG), Canadian Cattle Industry Association (CCIA), and a number of other industry-specific organizations that define standards proprietary to these organizations. These protocols usually do not support interoperability, but they are often available a lot sooner than the standards from the main organizations because of faster execution and relative simplicity.

Air Interface Protocols

Air interface protocols define the frequency at which the interrogator operates (keep in mind that the tag frequency must match the interrogator's frequency), the maximum power allowed for transmission by the interrogator, the modulation and demodulation schemes used by the interrogator to communicate with the tags, and the data rate of which the communication is capable. The air interface protocol also governs the interrogator-to-tag communications—such as commands for controlling a tag population, reading a tag, writing to a tag, or killing a tag—as well as tag-to-interrogator communication, such as responses from tags in the event of filtered read (that is, case tags responding to a call for only case tags) or in the event of a collision.

Interrogators that are expected to operate globally usually support a large number of air interfaces and data interfaces. Support for ISO, EPCglobal, AIAG, IntelliTag, and many other interfaces are used in such interrogators. Multi-protocol support is challenging, because the more protocols an interrogator scans for, the slower it reads. In high-speed applications, this can create issues that could be avoided simply by standardizing practice on a single protocol.

Data Protocols

Tag data protocols or data formats govern the tag memory and its layout, provide standards for data translation, and generally govern the meaning of the data as opposed to the way it is communicated, which is taken care of by the air interface protocols.

Dense Interrogator Mode

Dense interrogator mode (often called *dense reader mode*) allows for operation of multiple interrogators located within close proximity of each other without causing interrogator interference. Dense interrogator mode can also involve special allocations in which interrogators and tags are separated in their ability to communicate on various channels so that strong interrogator signals will not overpower weaker tag responses.

Interrogators that are EPC Gen 2 certified use dense interrogator mode when the number of interrogators present in an environment exceeds the number of available channels. In North America, dense interrogator mode is used if more than 50 interrogators are installed within a facility. In Europe, the mode is used if the number of operating interrogators is greater than 10 if they operate at maximum allowed power.

Dense interrogator mode allows for coordination of interrogators so that no two interrogators are transmitting at the exact same moment on exactly the same frequency and causing interference within one another's IZs. To do this, many interrogators perform frequency hopping and support a function known as *listen before talk* (LBT).

Frequency hopping employs a technology that forces an interrogator to change channels constantly within a given range of frequencies. For instance, a UHF interrogator licensed by the FCC for operation in North America must support hopping across 50 channels (each 500 kHz wide) between the frequencies of 902 to 928 MHz and spend no longer than 0.04 second in any channel during this rotation.

To perform LBT, interrogators use an antenna (often a dedicated antenna) to listen for the frequency on which the interrogator is about to transmit. If another interrogator is communicating in that channel, the interrogator will automatically switch to the next available channel and transmit there instead. LBT is often used with frequency hopping and is required in Europe for dense interrogator operations.

Other forms of dense interrogator mode include spectral allocation, time slice interrogator control, software synchronization, and hardware timing controls of interrogator platforms.

Spectral allocation designates a portion of the allotted frequency range for each type of communication. This allows tags to talk in different channels than interrogators, which prevents them from interfering with one another.

Exam Tip

Know that spectral allocation is the preferred method used by Generation 2 systems.

Time slice interrogator control is typically done with a back-end system controller that determines when each interrogator is allowed to communicate. In this method, each interrogator is provided a set period of time for communication, and it is then turned off to wait for its next turn while all the other interrogators communicate.

Software synchronization allows the interrogators to communicate with one another and determine which one can transmit at what time via software built into the interrogator's operating system.

Hardware timing will usually employ the use of triggering devices to allow interrogators to communicate for controlled periods of time. For instance, when an object breaks a light beam on the way into the IZ, the read cycle is started. As the object leaves the IZ, it breaks another beam that tells the interrogator to stop.

Tag Population Management

Interrogators support various commands and functions used for managing the tag population. They can address an entire group of tags to perform inventory of all tags in the area, as well as singulate each tag to access its memory and perform reading and writing operations.

The following are the most important interrogator commands:

* **Select** Used to determine which groups of tags will respond to further interrogation using the Inventory command (for example, this command can be used to isolate case tags from item tags). This command does not require any response from tags.

* **Inventory** Used to identify individual tags within a group.

* **Access** Used to communicate with individual tags and issue commands to them once they have been singulated.

* **Kill** After the tag has been accessed and a secure communication channel established, this command can be used to make the tag stop functioning. Killed tags will not respond to interrogation and cannot be resurrected.

* **Lock** Used to secure the contents of a tag. Once issued, it can prevent the tag from being read from or written to. Can also be used to lock individual tag memory banks.

Gen 2 interrogators support these interrogator commands and also tag management functions, such as AB symmetry and sessions.

AB symmetry replaced the Gen 1 technique of "putting tags to sleep" after being interrogated. AB symmetry allows tags to be flagged with an identifier to ease counting. If a tag that was flagged as an *A* gets counted, it then becomes flagged with a *B* until another round of inventory, when it returns to its default position A. This technique eliminates problems with tags that cannot "wake up" or are slow to wake up for the next round of inventory.

Gen 2 tags support four *sessions*. This function is used when more interrogators or groups of interrogators interrogate the same group of tags. A tag would use each session to communicate with one interrogator or group of interrogators. This way, if a tag communicates with one interrogator and has been flagged with a *B*, for instance, a second interrogator communicating with the tag will use a second session with its own *A* and *B* flags. This will prevent the interrogators from interfering with each other's inventory rounds and avoid confusion that may cause an interrogator to count a tag twice.

Anti-collision

The various types of anti-collision methods can be reduced to two basic types: *deterministic* and *probabilistic*. An example of deterministic is binary tree algorithm or tree walking algorithm. An example of probabilistic is the ALOHA anti-collision algorithm.

The deterministic algorithm works by asking for bits on the tag ID, and only tags with matching IDs respond. It starts by asking for the first numbers of the tag until it gets matches for tags; then it continues to ask for additional characters until all tags within the region are found. This method is slow but ultimately leads to fewer collisions and a more accurate search for tags in the IZ.

In a probabilistic method, tags respond at randomly generated times. If a collision occurs, colliding tags will have to identify themselves again after waiting a random period of time based on a random number the tag selects. This process will eventually isolate and identify all the tags in the IZ but is prone to collisions since it is possible that tags could choose times too close to one another.

Interrogation Zone Configuration

Objective 4.5

To make an interrogator function properly, you must configure it properly using settings that differ by the type of the interrogator as well as the manufacturer. Following are the most common settings you may encounter.

- **Output power**   One of the most important settings is the transmit power level of an interrogator. The power level is usually set so that interrogators do not interfere with IZs installed parallel to one another. Power levels will be usually provided in watts (W), milliwatts (mW), decibels (dB) or as a percentage, where 100 percent will equal to 1 W (or the maximum allowed power for that region of the world). In certain situations, you may be able to adjust both the reading power and writing power.

- **Antenna settings**   You can combine antennas into groups within the configuration of an interrogator. This allows multiple antennas to act as a single device, pushing more power into the IZ to flood the area with a signal and ensure that all tags in that area are read. In many cases, you can also choose antenna sequencing, which allows multiple antennas to operate in a particular order.

- **Retries**   You can set up the number of times an interrogator will attempt to find new tags in the IZ before reporting its findings. This is known as *retries*.

- **Filtering**   This feature allows the interrogator to communicate only with the portion of the tag population it is looking for. For instance, a dock door interrogator may filter out all case tags and communicate results from pallet tags only.

- **Polling**   Interrogators can scan for tags either continuously, in preset intervals, or on demand.

- **Modes of reading**   Sometimes you can choose from various modes of reading that are already preset to ensure a specific interrogator performance. You can encounter a conveyor mode, self-mode, or rapid read mode, and although manufacturers use different terms to refer to these modes, they all are basically the same for all interrogators. When in conveyor mode, the interrogator reads in fast intervals with a low number of retries; in this mode, usually few tags are in the IZ but they need to be read quickly, since the tag is in view of the antenna for only a short time. Inventory mode is more of a transitional mode. The interrogator will report when it sees a change; it will see a large number or a given number of tags for a specific amount of time, and when that number changes, the interrogator instantly generates a report to the back-end system to indicate the loss or gain of a new tag identity.

Handling Interference

Interference is mitigated by the environment and the modes in which the interrogators operate. *Single interrogator environments* obviously indicate one interrogator is installed in a particular facility. This interrogator can broadcast on whatever channels it finds appropriate within its configuration at any given time. The interrogator will not encounter interference from other interrogators, but it may encounter interference from the environment and from its own multipath interference caused by reflections of its own or other signal(s). This type of interference can be handled by proper shielding and by adjusting the signal strength and antenna gain.

If the number of simultaneously operating interrogators is smaller than the number of available channels, a *multiple interrogator environment* results. In this type of environment, the interrogators can interfere with each other, but you can resolve this by assigning specific channels to specific interrogators or by time division.

In a *dense interrogator environment*, the number of interrogators operating is greater than the number of available channels. As discussed earlier, certified interrogators will incorporate schemes identified in the Gen 2 specification to minimize mutual interference.

Exam Tip

Know that to handle a dense interrogator environment successfully, it is essential that all interrogators operate in dense interrogator mode. If one interrogator does not operate in this mode, the whole scheme will not work.

System Optimization

You must consider the speed and direction of the tagged assets/goods moving through the IZ when you are installing and configuring the IZ. Direction can be determined by using multiple input sensors, such as light break sensors, at the beginning and the end of the IZ. The order in which the sensors are triggered will indicate which direction the goods are traveling through the IZ. Some antennas in development can sense the direction of tag movement. These are essentially *active antenna arrays*, which will sweep a coverage area using multiple antennas embedded within one another in a certain sequence. These antennas, coupled with the appropriate interrogator, can not only indicate direction of travel, but also eliminate weak areas in the coverage of the primary lobe.

Local Lingo
Dwell time   The time in which tags are within range of an antenna's beam.

The speed of the items traveling through the IZ will ultimately determine how many tags can be successfully read during their dwell time. If an interrogator has a maximum read ability of 500 tags per second, for example, and due to the coverage of the antennas and the speed of the goods, tags are in the interrogation field for only two thirds of a second, your interrogator will read only 330 tags—two thirds of the total tag population. These values come from the combination of the constraint of the abilities of the interrogator and the speed the tags travel through the zone. If you want to extend the dwell time as much as possible, you should angle the antennas 25 to 45 degrees into or away from the path of travel. This will make the beam wider along the path of travel for the goods, thereby increasing the time the tags spend in the read field.

Travel Assistance
For more information on antenna positioning, see Chapter 7.

Angling the antennas will also help with handling the reflections in the IZ caused, for instance, by metallic objects.

Travel Advisory

If you know that you will be reading tags placed on metallic items or materials, you should install multiple antennas, each slightly angled, so that at least one antenna captures the backscattered signal from the tag without interference from reflected waves.

CHECKPOINT

✔**Objective 4.1: Interrogator Functionality**   Interrogators read and write data to and from tags. Interrogators are responsible for remotely powering the tags (in passive and semi-passive systems), establishing the bidirectional data flow between the tags and themselves, and converting analog to digital as well as digital to analog signals.

Interrogators can also run attached peripheral devices such as light stacks or alerting horns, which provide feedback on conditions within the system, and triggering devices, which control when the interrogators are turned on and off. Peripherals are usually attached to the I/O ports of the interrogators.

Most interrogators, depending on their type, include some of the following communication ports: RS-232, RS-245, or RS-422 serial ports; USB; Bluetooth and 802.11 wireless network interfaces; LAN ports; and general-purpose I/O ports (GPIOs).

Firmware upgrades on interrogators are performed to enhance performance, support new features, or resolve bugs in the existing platform. They can be downloaded from the manufacturer's site and deployed to the interrogators in a variety of methods, ranging from FTP, HTTP, or automated push, depending on the manufacturer and middleware used in the installation.

The interrogator's GUI is typically accessible through a web interface using any commercial Internet browser or through a simple application program. Typical features of a GUI include the ability to read and write tags; make network configuration settings, antenna settings, and power settings; configure an I/O port; and upgrade the firmware.

✔**Objective 4.2: Interrogator Types**   Interrogators are available as fixed or mobile devices that can be handhelds or vehicle mounted.

Handheld interrogators are often used for low-volume reading or writing in exception processing, quality assurance, or mobile shipping-receiving units. Handheld interrogators can be tethered, receiving power and accomplishing data transmission via a nearby base unit using an extensible cord. Other types of handhelds can be wireless that depend on wireless networks or Bluetooth communications for data transfer, limiting operational time to battery life between charges. Newer renditions of handheld interrogators are incorporated in cellular phones and PDAs.

Vehicle-mounted interrogators are usually integrated into material-handling devices such as forklifts, paper trucks, cargo trucks, and pallet jacks. These interrogators usually have a special shape for easier installation on the vehicle and a rugged design to survive vibrations and other environmental conditions.

✔**Objective 4.3: Interrogator Antennas**   Antennas can be linearly or circularly polarized. Linearly polarized antennas propagate only on one plane, while circular antennas propagate on all planes. Circular antennas are less efficient because they have to divide the signal between multiple planes, while linear antennas can focus on only one plane.

Antennas can be bistatic or monostatic, which is determined by an interrogator. A monostatic interrogator uses a single antenna to transmit and receive and requires a circulator, which lives inside the interrogator and turns the antenna from transmitting to receiving function in a rapid manner. A bistatic interrogator uses two antennas, each dedicated either for transmitting or receiving. Bistatic interrogators usually have twice as many antenna ports as monostatic interrogators.

✔**Objective 4.4: Interrogator Operation**   Dense interrogator mode allows for operation of multiple interrogators within a close vicinity of each other without causing interrogator interference. Dense interrogator mode has several variations, such as spectral allocation, time slice interrogator control, software synchronization, and hardware timing. When installing multiple interrogators, you can help prevent interference by shielding them and making sure that the antenna footprints from adjacent zones do not overlap.

Interrogators that are EPCglobal Gen 2 certified use dense interrogator mode when the number of interrogators present in an environment exceeds the number of available channels. In North America, dense interrogator mode is used if more than 50 interrogators are installed within a facility. In Europe, the number of operating interrogators must be larger than 10 if they operate at maximum allowed power. In a dense interrogator mode, many interrogators perform frequency hopping and support the listen before talk (LBT) algorithm.

In the United States, a UHF interrogator licensed by the FCC for operation in North America must support hopping across 50 channels (each 500 kHz wide) within the frequencies of 902 to 928 MHz and spend no longer than 0.04 second in any channel during this rotation. LBT is a requirement of operation in Europe in the band of 865 to 868 MHz. Operation at 869.4 to 869.65 MHz requires duty cycling.

Two major types of anti-collision protocols are used: deterministic and probabilistic. The deterministic algorithm works by asking for bits on the tag ID, and only tags with matching IDs respond. This method is slow but ultimately will lead to fewer collisions and a more accurate search for tags in the IZ. In a probabilistic method, tags respond at randomly generated times. If a collision occurs, both tags will have to identify themselves again after waiting another random time interval. This process will eventually isolate and identify all the tags in the IZ but is prone to collisions. The Q algorithm is used during anti-collision by Gen 2 tags.

Interrogators perform tag population management using commands such as Select, Inventory, and Access (including Read, Write, Lock, and Kill).

✔**Objective 4.5: Interrogation Zone Configuration**   Several settings must be configured in an IZ: output power, antenna settings, retries, filtering, polling, and modes of reading such as a conveyor mode, self-mode, or rapid read mode. These modes are preset configurations for specific types of applications.

To achieve the highest read rates, the interrogator should be set only for one air interface protocol. Use dense interrogator mode only if necessary because it slows down interrogator performance; however, dense interrogator mode will ensure that the tags are read successfully.

The system should be set to interrogate only when needed using triggering devices such as sensors. The transmitted power must be set according to the desired radiation pattern and read distance to achieve successful reads and reduce interference. Using antennas with higher gain will achieve longer read ranges but a narrower read field, while antennas with lower gain will achieve shorter read ranges and a wider read field. Angling the antennas 20 to 45 degrees toward the direction of tag travel or following the direction of travel helps increase dwell time and therefore reads and may help with avoiding interference. Angling the antennas will also help with handling the reflections in the IZ caused, for instance, by metallic objects.

REVIEW QUESTIONS

1. An interrogator mounted on a forklift is required to send real-time data to a host computer in the warehouse. Which of the following is the best way to handle this situation?

   A. Get a smart interrogator that collects data, and then load this data to the host at the end of the shift.

   B. Connect an Ethernet cable to the interrogator on the forklift and connect another end of the cable to the server in the computer room.

   C. Install a computer on a forklift, connect the interrogator to it via serial cable, and let the computer handle the data transfer.

   D. Get an interrogator with a wireless LAN port, and send data to the host.

2. Tags mounted on metal car parts are passing through the IZ on a conveyor. What is the best way to reduce the interference due to reflections caused by the metal surfaces in the IZ?

   A. Increase the power output from the interrogator.

   B. Install multiple antennas at an angle to each other.

   C. Install a vertically polarized antenna.

   D. Decrease the power output of the interrogator.

3. The protocol used for control of the interaction between the tag and the interrogator is known as the

   A. Air interface protocol

   B. Data interface protocol

   C. A and B

   D. Anti-collision

4. A distribution center stacks pallets on an open floor divided into storage areas. Each area has a passive UHF tag installed in the floor for tracking purposes. Which of the following options would be the best way to install the tag when forklifts are in use?

   A. Hang the tag from the ceiling.

   B. Stick the tag directly to the floor.

   C. Place the tag in a shallow void covered with a polycarbonate layer.

   D. Embed the tag flush into the concrete ensuring that it is not covered.

5. A warehouse in Berlin has 25 dock doors with passive UHF Gen 2 interrogators. The warehouse operates 24 hours, 7 days a week, continuously receiving and shipping items through the dock doors. Which of the following modes of operation should be used for all the interrogators?

   A. Single mode

   B. Multi-interrogator mode

   C. Dense interrogator mode

   D. Many interrogator mode

6. When a UHF interrogator has one tag in its IZ, it can read that tag a maximum of 100 times in 1 second. What would happen if 98 tags are simultaneously passing through the IZ in 1 second?

   A. All tags will be read.

   B. Two tags will be read twice.

   C. Some tags will not be read.

   D. No tags will be read.

7. What is the function of an anti-collision algorithm in a passive UHF system?

   A. It allows the interrogator to interrogate all the tags in its zone (one tag at a time).

   B. It allows multiple tags to be interrogated at the same time.

   C. It allows multiple interrogators to interrogate the same tag at the same time.

   D. It allows multiple interrogators to operate in dense interrogator mode.

8. An antenna port on a monostatic interrogator allows for what?

   A. Either only receive or only transmit signals from and to tags

   B. Both transmit signals to and receive signals from tags

   C. Listen before talk function

   D. Block electrostatic charge

9. Which of the following is not available on handheld interrogators?

   A. Multiple antenna ports

   B. Ethernet ports

   C. Wireless LAN ports

   D. Serial ports

10. According to the EPCglobal Class 1 Gen 2 protocol, interrogators operating in dense interrogator environments are separated in what way?

   A. Temporally

   B. Logically

   C. Physically

   D. Spectrally

REVIEW ANSWERS

1. **D**   When an interrogator mounted on a forklift is required to send real-time data to a host computer in the warehouse, you should get an interrogator capable of wireless communication and send data to the host.

2. **B**   The best way to reduce interference due to reflections caused by the metal is to install multiple antennas at an angle to each other.

3. **A**   The protocol used to control of the interaction between the tag and the interrogator is known as the air interface protocol.

4. **C**   When installing a passive UHF tag in the floor for tracking purposes in the distribution center, it is best to place the tag in a shallow void covered with a polycarbonate layer.

5. **C**   In a warehouse in Berlin with 25 dock doors with passive UHF Gen 2 interrogators, you must use a dense interrogator mode, because the number of channels in Europe with maximum allowed transmitted power of 2 watts ERP is 10.

6. **C**   If 98 tags are simultaneously passing through the IZ in 1 second, collisions may occur; therefore, it is likely that some tags will not be read.

7. **A**   The function of an anti-collision algorithm in a passive UHF system is to allow the interrogator to interrogate all the tags in its zone (one tag at a time).

8. **B**   An antenna port on a monostatic interrogator allows signals to be both transmitted to and received from tags.

9. **A**   Multiple antenna ports are not available on handheld interrogators. Handhelds usually have only one antenna integrated in the unit without the ability to connect more antennas.

10. **D**   According to the EPCglobal Class 1 Gen 2 protocol, interrogators operating in dense interrogator environments are separated spectrally.

RFID Peripherals

	NEWBIE	SOME EXPERIENCE	EXPERT
ETA	4 hours	2 hours	45 minutes

Every RFID implementation needs peripheral devices such as RFID-enabled printers, RFID-enabled automated label applicators (also known as print and apply devices), as well as feedback components such as light stacks, horns, LED lights or LCD displays, and triggering devices such as light break sensors. These devices enhance the capabilities of an RFID system, control I/O devices, control system timing functions, and provide on/off capabilities for various business applications. Peripherals also improve human interactions with RFID hardware by assisting automatic data collection and increasing the safety of the system.

 Installation and Configuration of RFID Printers

RFID printers in an automated RFID system often utilize existing print points within a manufacturing or distribution supply chain. Because many of the current Auto-ID systems use barcoding, the conversion to RFID systems will be an incremental change. The first step in enabling your current barcode printing and application points with RFID will be to obtain RFID-enabled label printers.

RFID-enabled label printers represent a tagging method that can accommodate both types of Auto-ID systems, RFID and barcode, in a single unit. The system uses a regular barcode label with an *RFID inlay* embedded in the back of the label. This is called a *smart label* and will typically have human-readable information printed on the front of the label. The addition of the RFID tag enables non–line-of-sight identification, which allows for many automated functions and/or serialized inventory.

Using a smart label printer is the easiest way to handle RFID integration, because it is easily incorporated with the existing business process, data structures, and networks associated with it. The smart label printer replaces the existing printer, keeping the label application process identical. The same print server and network connections used for the old printer can be used with the new one.

The labels can be printed and applied immediately prior to shipment to downstream clients. This application, typically called *slap and ship*, is one of the fastest and easiest ways to comply with any customer mandates. However, slap and ship does not provide any of the benefits of RFID for the user, with the exception of verification of shipping (Figure 5-1 shows an interrogation portal

FIGURE 5.1   Interrogation portal

used for verification). This is sufficient for a company shipping limited amounts of stock keeping units (SKUs). However, when companies reach a point at which costs of the manual labor used to apply labels onto products prior to shipping outweigh the costs and benefits of tagging all of their products during manufacturing process in an automated manner, slap and ship systems will become obsolete. If you plan properly, you can reuse much of this equipment in the larger, full-scale implementation.

Label Selection

To use an RFID-enabled label printer, you must first perform a case analysis to determine what tags are correct for use with your application. The case analysis will include a detailed assessment of the contents of the package and its effect on RF, the package material and its effect on RF, the package surface characteristics for selection of the proper adhesive, the packaging design, and the method of attachment to be used for tagging the cases and products. Figure 5-2 shows a smart label, which is the most common type of media for passive-RFID tagging.

FIGURE 5.2 A smart label

The label should be large enough to display all of the existing printed information as well as any additional information required for the RFID tag embedded in the back of the label, such as the Electronic Product Code (EPC) logo to indicate the use of an EPC tag within the label.

The size should also be appropriate for the boxes to which the labels are applied. You must also consider the compatibility of the label adhesive with the boxes, the method of attachment, the environment in which the tagged assets will be moving, as well as the requirements of the label applicator. The RFID tag embedded in the label should meet pre-existing RFID customer requirements. Some of these requirements can include the standard that the customer uses for reading within the facility, the frequency of operation for the tag, as well as the amount of memory on the tag, read/write capabilities, and the tag's ability to contain the data required.

The RFID tag embedded in the label should be appropriate for the items inside the box. Special tags may be needed to tag difficult-to-read materials such as water, metal, and most aqueous liquids, as well as packaging or product with high carbon or salt contents. All of these materials can affect the RF signals coming from the interrogator as they attempt to energize the tag.

You must also consider the compatibility of labels with your label applicator, including the dimensions of the label and whether or not they fall within the acceptable parameters for the printer. Two main types of label media formats are used: *media roll* or *fan-fold*.

The label surface must be appropriate for the method of printing used with the particular printer. Two types of print methods are used for both industrial barcode printers and RFID printers: *thermal transfer*, which uses heat to transfer ink from a separate ribbon to the label, and *direct thermal*, which uses special media in the form of a treated label stock that turns black when heated. Pros and cons exist for both of these methods. For long-term application where the labels will have a lengthy lifecycle in the supply chain, thermal transfer printers should be used as the labels do not fade with exposure to sun or heat. These printers will typically print with a higher DPI (dot per inch), or resolution, on the label. The advantages of a direct thermal printer are a simpler design and therefore easier loading of the media, no ribbon, and easier maintenance for the printer due to fewer components that could break. However, the labels tend to be more expensive and fade quicker when exposed to heat.

> ### Exam Tip
> Be aware that for labels that will be subjected to heat, a thermal transfer printer should be used.

When choosing a tag and a printer, you must use the same protocol and frequency as well as proper inlay alignment in a label. The RFID tag must be positioned so it aligns correctly with the encoder antenna so that it can be read or written to. This positioning changes from one printer to the next, as many printers have their antennas placed at different locations within the print pad. Each label will be specific to a particular type of printer for which it is designed and the inlay position in the label should match the specifications provided by the printer manufacturer. If the inlay position does not match the required specifications, you will not be able to use a default mode and you will need to run a calibration process to make sure the printer determines the inlay's position. You can see various tag alignments in the smart label in Figure 5-3.

When a tag inside a label fails to operate within the specification of that tag, it is called a *quiet tag*. Quiet tags can be a major concern for label makers as well as users. Tag quality and performance need to be consistent. If it takes more energy to light one tag than the others, this tag is not going to perform as well in your

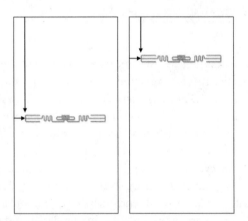

FIGURE 5.3   RFID tag alignment inside a smart label

interrogation zone (IZ). Performance of your tags must be reliable and must have repeatable results that you can count on, in order to properly assess the inventory moving throughout your facility.

Most RFID-enabled label printers have a way of verifying the function of the tag in the label and assessing its readiness for use. These printers will determine the response of the tag, and if the tag does not fall within the manufacturer's suggesting quality guidelines, the tag will be rejected and a special pattern will be printed on the front of the label, indicating it should not be used. The printer will then generate the next available label on the print line.

The tag within a smart label contains delicate electronic components and must be handled just as you would handle computer components. Take care to avoid subjecting tags to electrostatic discharge (ESD), crushing, dropping, and exposing to moisture and extreme temperatures. You must consider all of these aspects when choosing proper storage locations and handling procedures before you purchase and use a smart label.

> **Travel Advisory**
>
> Most RFID-enabled printers have validation and error recovery built in, and the command structure to control the printer is similar to existing barcode printers, with the addition of RFID control commands to enable the encoding and verification of smart labels.

Types of RFID Printers

Printers come in several varieties, including the following:

- Plain label printers, which can use pre-encoded smart labels and print information on them
- Printer encoders that print information on the label and then encode the tag with an EPC number or with another relevant data representation for the product to which it is attached
- Printer encoder and applicators, which print information on the label, encoded with the identifying number, encode the RFID tag with the proper identification, and then automatically apply the label to the box

The following photos show several types of RFID-enabled printers. Many of these are designed in industrial-strength boxes typically made of metal for use in heavy industrial environments where exposure to dust, temperature, high humidity, and other extremes are likely.

Travel Advisory

As with any technology, the current landscape will change in six months. Protect your investment as much as possible by buying print devices (this is valid for all devices) that are firmware upgradeable, for both the printer and the RFID unit embedded within it. Upgrading firmware to support a new protocol is much less expensive than buying a completely new printer.

Take care to follow the proper paper path for the media when loading a printer, as this is imperative to the printer's success when attempting to encode smart labels. The print path is typically shown on the backboard of the printer when you open the side cover, as shown in Figure 5-4. You must align the media coming off the spool with the line painted on the backboard of the printer as you feed it into the print head.

Tag Encoding

The printer must isolate the individual tag to which it is writing during the print process. For this reason, RFID printers have integrated a very low power reader with an antenna designed to read tags that are in close proximity. This requires proper alignment of the RFID tag within the label and within the print pad.

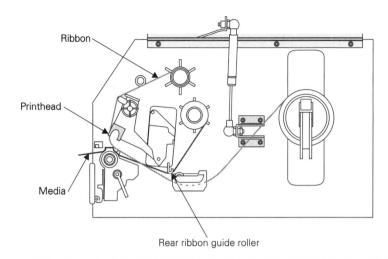

Ribbon

Printhead

Media

Rear ribbon guide roller

FIGURE 5.4   Inside of an RFID printer

Exam Tip

Remember that writing to a tag requires more time and power than reading from a tag, so this means that a writing application will be a lot slower than simply reading the tag and applying it to the box.

Local Lingo

Writing to a tag   Also called tag encoding. The encoding or writing is done by an RFID printer. Data can also be encoded or written to a tag by RFID interrogators

The following sequence of events takes place to print and encode a label:

1. The printer receives a command and data from the host computer.

2. The label is positioned so that the reader can check for null data on the tag. Null data is placed in a tag by the tag manufacturer to indicate that

the tag has passed the manufacturer's quality control. A quiet tag that was pre-identified at the manufacturer will not have null data written; therefore, this tag will be automatically rejected. The printer does not attempt to program this tag and does not self-test the tag for any signal strength but simply rejects it.

3. If the tag passes a validity check, the printer will write data to the tag and then verify the data it has just written.

4. If the verification fails or the tag does not respond with the appropriate signal strength, the tag is marked with a special void mark; it is then rejected, and the process is repeated on the next label.

5. Once the tag is successfully encoded, the barcode and other information is printed onto the label and an acknowledgement is sent back to the host. This acknowledgement contains the serial number that was just written to the RFID tag, as well as a successful print signal.

6. The printed and encoded label is now presented to the user for application to the package.

Printer Setup

Proper printer setup is extremely important for successful printing. The most important setup parameters for RFID enabled barcode printers are these:

- **Label dimensions**   Typical sizes for these are 4×1, 4×2, 4×4, and 4×6 inches.
- **The distance between two consecutive labels**   This measurement combined with the label length determines how far the printer will advance to print/encode the next label.

- **The location of the inlay within the label** Some RFID enabled printers offer the ability to move the antenna along the length of the label. This is usually specified when the labels are ordered to ensure compatibility with a specific manufacturer's printer.
- **The inlay type** Some printers have settings that allow you to set the specific manufacturer and model of the inlay being used in the printer.
- **The inlay protocol** If a generic tag size is chosen without specifying the make and model of the inlay, a tag protocol will need to be defined. This lets the encoder know what air and data protocol to use when communicating with the tag.
- **The dimensions for various print zones** This setting is typically found in the software driving the printer. When formatting the label print area, this portion can be used to set up the printer to avoid printing on the portion of the label containing the tag. This area can cause artifacts on the printed label that can make portions of printed information unreadable.
- **The feed rate for the printer** This is used to control how fast the printer moves the labels through the print area.

Printer Installation

During printer installation, safety must be a prime consideration. Surge protection for the electrical supply and pressure relief for the pneumatic supply used with print applicators are necessary precautions.

Follow these steps for printer installation:

1. Identify a proper location. When looking for the appropriate location for a printer, ensure that it can be reached easily and is on a dry, level surface. Operators need to have clear access to the front of the printer to obtain printed labels. Additionally, adequate space should be available to raise the cover of the printer for changing media and for maintenance. The printer must be within cabling distance of power and data connections. Finally, ensure that the location has adequate environmental controls and ventilation to keep the printer within its operational parameters as specified by the manufacturer.

2. Connect the printer to a consistent, grounded power source. The printer should come with a grounded power cord, which you must

connect to a surge protected and reliable power source. If unstable power conditions are common at the site, install an uninterruptible power supply (UPS) in between the printer and the power outlet.

3. Connect the communications cabling and configure the device for the network it is being connected to. There are many different types of communications used by print devices. These include:

- **Ethernet**   This communication method will commonly employ a CAT5 cable to connect to the back-end network. The most critical settings for this type of communication are IP address, subnet mask, and default gateway.

- **Wireless Ethernet**   Wireless networks require configuration of all settings found in Ethernet networks, with the addition of the appropriate wireless network credentials. These typically include the service set identifier (SSID) and wired equivalent privacy (WEP) key for security.

- **Serial networks**   Serial ports vary from one type to the next, but common types are RS-485, RS-422, and RS-232. Settings will include baud rate, bits per second, parity, stop bits, and flow control.

4. Load the media and run a test print. Refer to the printer manual for instructions on how to run a test print, since this varies by manufacturer. Most test labels will print a status page showing information about the printer settings so you can verify all the information you just configured. If everything has been configured correctly and the test print is successful, you should be able to generate a label using the back-end system.

Objective 5.2   Automatic Label Applicators

Automatic label applicators are used for integration of RFID into an automated process. They can typically print on the face of the label and encode the RFID tag that is embedded within it. Then this label is placed in a predetermined location on objects as they pass. Speeds vary according to the technology used and the application method employed.

There are two main types of RFID encoder printer applicators or automated printing code and apply devices:

- Pneumatic piston label applicators (Figure 5-5), also called tamp or blow applicators, have a pneumatic tamp head that pushes labels onto the object at a precise moment to ensure specific placement with every label.
- Wipe on label applicators (Figure 5-6) simply present the tag adhesive side toward the package at the proper time in the manufacturing line for the adhesive side to grab the side of the object to adhere the media in the same spot on the side of the box every time.

FIGURE 5.5 Pneumatic piston label applicator

Wipe on label applicator

Setup and configuration of automatic label applicators involves the same in-stallation procedures used for a smart label printer, with the addition of a few extra steps:

- Immediately after connecting the unit to power, connect it to an air supply that meets the manufacturer's specifications.
- During the configuration portion of the installation, special adjustments may need to be made to reflect the proper timing of the application process, position and orientation of the label, stroke of the pneumatic assembly, and pressure to be exerted on the label. These settings are specific to the manufacturer, so be sure to consult the manual while installing the unit.

 **Objective 5.3** **Feedback Systems**

Feedback systems are used to alert operators to conditions that equipment has been or is currently experiencing. The conditions indicated are configured specifically to an application. Many common events that would cause an alert are successful or unsuccessful tag read events, successful or unsuccessful tag write events, improper flow of materials, or items being mishandled. Because a limitless number of events could cause an alert, the parameters that define when an alert occurs must be carefully planned and designed.

The most commonly used feedback devices are the following:

- **Lights** Lights can be used as part of a control panel, an overhead lighting alert, or a light stack. They are used in places where visibility is ensured and may be necessary due to high amounts of noise in the area that could render an audible alert ineffective.

  Normally, the lights are connected to a Programmable Logic Controller (PLC) or a general-purpose I/O (GPIO) port of an RFID interrogator. Events and conditions can trigger various lights depending on the programming of the device. For instance, a green light usually means a successful read, while a red light indicates error conditions.

- **LCD displays** LCD displays can be simple LCD panels that are capable of a limited amount of characters or LCD computer displays that can display large amounts of variable data and graphics. These are used when more information is required than is practical through simple notification systems.

- **Horns** Horns are used for environments in which a bulb or screen may be difficult to see for a variety of reasons: from being outdoors in the sun to the fact that the location of the indicators is not within sight of the operator during regular procedures. Horns and buzzers are typically loud and obnoxious to ensure prompt reaction to facilitate correction of an error condition.
- **Divert gates** During conveyance, exception triggers can cause automatic diversion of objects to automated handling of these events.
- **Feedback through the software** Software components can monitor sensors along your material handling process and notify when events occur. The actions taken by the software could be simple database logging, all the way to advanced automated exception handling.
- **Cell phone alerts** When specific conditions arise and have not been addressed, an escalation path can be followed for notification. For instance, local alerts are set, such as light stacks or horns, and if these are not acted upon, the system can escalate to an e-mail alert. If there is

still no action, a Short Message Service (SMS) message can be sent to one or many cell phones to be handled by appropriate personnel.

Travel Advisory

Make certain that if you enable the system to send an alert, a positive reaction is required, such as a button press or a reintroduction of the product to the IZ. This way, the system can record with certainty that the alert was received and processed, and that actions were taken as a result.

Objective 5.4 Triggering Devices

Triggering devices are used to detect the presence of an object in an IZ. Every well-designed RFID system should have triggering devices to ensure the longevity of the equipment, the safety of the system, and the best possible environment in which the system can operate. Without these sensors, interrogators are left to poll constantly, which degrades the equipment over time, floods the area surrounding the IZ with high levels of ambient electromagnetic noise, and can potentially create safety issues due to lengthy exposure to RF for anyone working in the immediate proximity of the IZ.

- **Light break sensor** Light break sensors are commonly employed in manufacturing conveyor applications, though they can be used in many other scenarios as well. They consist of two components, an emitter and a receiver. The emitter creates and sends a light beam across the travel path for the object you are trying to detect. The receiver is placed directly across from the emitter, and it recognizes the light beam. When the beam is broken, this indicates that an object has entered the area. The receiver then sends a signal to a controller that will initiate whatever action is required based on the settings.

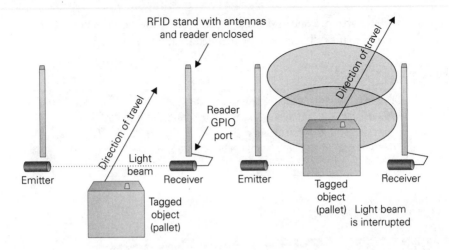

- **Proximity sensor** Proximity sensors typically use ultrasonic waves to detect the presence of an object in the area. An object passing close to the proximity sensor will reflect the ultrasonic waves back at the unit, which can then derive the approximate distance based on the time for the travel of the waves.

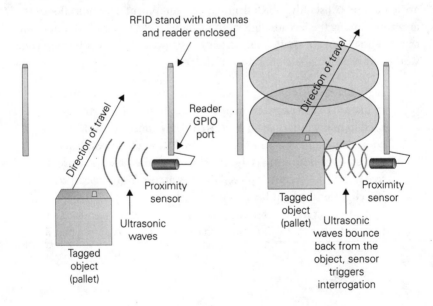

- **Other types** Other types of triggers can be employed, based on the specific needs and inputs/outputs of a particular system. Software algorithms that calculate the location of an object based on the speed it is traveling and the last station it passed through, weight sensors to detect an object that is directly on top of it, pressure sensors connected to hydraulics of material handling devices to indicate an object is being moved with it, and motion detectors similar to those used in home security systems can all be used to trigger events within your system.

Real Time Location Systems

Objective 5.5

Real time location systems (RTLSs) are employed to track high-value or mission-critical objects. As it applies to RFID, this type of system employs an active tag that will assist a back-end system in calculating its position relative to known points in the supply chain.

The functionality and methods used to determine the location of tagged goods is similar to the way a GPS (global positioning system) works. However, the known points being measured from are within a facility or campus, and the range of the transponder attached to the tracked asset is usually measured in meters, most with a range of 300 to 600 meters, rather than miles or kilometers. RTLSs also do not employ a satellite and can be used inside buildings, while GPS works only outside, where the GPS receiver has access to a satellite.

Several types of RTLS systems are in use:

- **Differential Time of Arrival (DTOA)** DTOA systems employ a tag attached to an object that beacons on a given interval or is awaked upon an event and then beacons. When the tag transmits, receivers within range of the tag time stamp when they receive that particular ID. This information is then passed to a location processor that measures the difference between the receipt times of at least three receivers, which allows the processor to "triangulate" the position of the tag, as depicted in Figure 5-7. Currently, this is the most accurate location technology available today.

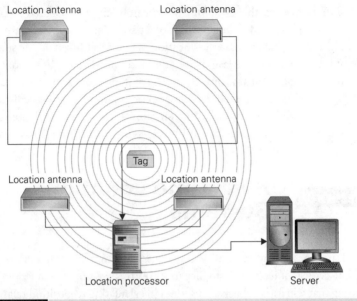

FIGURE 5.7   RTLS based on DTOA

- **Received Signal Strength Indicator (RSSI)**   RSSI systems are used commonly in 802.11 wireless networks. An 802.11 wireless device becomes the tag for this type of system. Based on signal strength, the systems can determine to which receiver the transmitter is closest and then provide an approximate location for it. The accuracy of this type of system will be determined by the granularity with which the receivers were installed. More accuracy is a direct reflection of installed cost. The working principle of this type of system is showed in Figure 5-8.

- **Other types of systems:**

  - **Time-of-flight ranging systems**   (Figure 5-9) These systems wake up tags on a given frequency, and then measure how long it takes the tag to respond on a different frequency, which gives the system an approximate distance for the tag from the antenna.

  - **Amplitude triangulation systems**   (Figure 5-10) These systems use the signal strength of tags received by multiple receivers to determine an approximate location for the tag. The processor

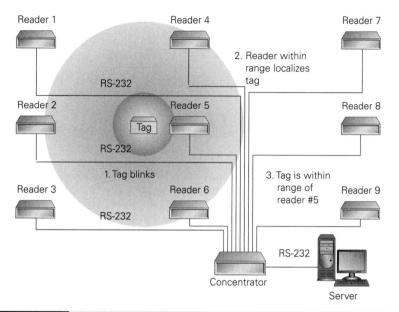

FIGURE 5.8 RTLS based on RSSI

behind the receivers compares the signal strength from multiple
receivers, and determines that the tag is between three of them
based on receive strength.

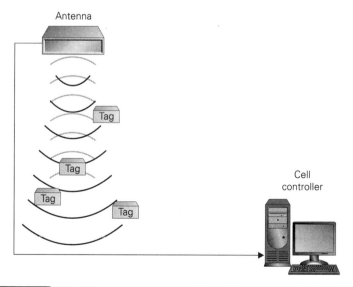

FIGURE 5.9 Time-of-flight ranging system

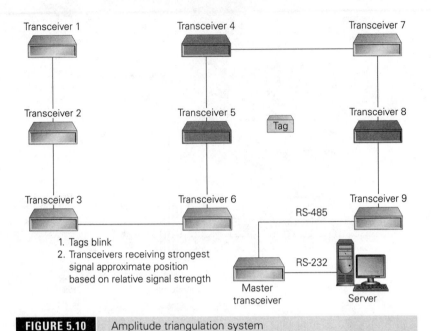

FIGURE 5.10   Amplitude triangulation system

Local Lingo

Triangulation   Triangulation is based on a measurement of distance in time. The speed at which a radio wave travels is absolute. Therefore, a triangulation system compares how long it takes a signal to reach from the tag to each of three receivers, and based on knowledge of relations of angles and sides in a triangle, the location of this tag can be determined, or "triangulated."

CHECKPOINT

✔**Objective 5.1: Installation and Configuration of RFID Printers**   RFID printers encode the smart label (a label with embedded RFID inlay) and print information on the surface of the label. Some printers are combined with automatic label applicators, with which the label after encoding and printing is automatically applied to an object. All RFID printers include an RFID interrogator inside to encode and verify the smart label.

The first step in printer installation is to identify a proper location for the printer, which must be easily reached and located on a dry, level surface to ensure access to the printer to obtain printed labels and to raise the cover of the printer for changing media and perform maintenance. The printer must be within cabling distance of power and data connections and have adequate environmental controls and ventilation.

You must connect the printer to a consistent, grounded power source and then to the communications cabling and configure the device for the network to which it is being connected. Printers use Ethernet, Wireless Ethernet, serial, and other types of communications. Lastly, you should load the media and run a test print. When configuring the printer, you must set up label dimensions, the distance between two consecutive labels, the location of the inlay within the label, the inlay type, the inlay protocol, dimensions for various print zones, and the feed rate for the printer.

✔**Objective 5.2: Automatic Label Applicators**   Automated label applicators are used for integration of RFID into an automated process. They can typically print on the face of the label and encode the RFID tag that is embedded within it, and then place this label in a predetermined location on objects as they pass. Speeds of the process vary according to the technology used and the application method employed. Two main types of label applicators are used: wipe-on label applicator, which presents the tag adhesive side toward the package in the manufacturing line so the adhesive side grabs the side of the object to adhere the media in the same spot on the side of every box, and pneumatic piston label applicators, also called tamp or blow applicators, which have a pneumatic tamp head that pushes labels onto the object at a precise moment to ensure specific placement with every label.

✔**Objective 5.3: Feedback Systems**   Feedback systems are used to alert operators to conditions that equipment has experienced or is currently experiencing. Most common feedback devices and methods are light stacks, horns, LCD displays, divert gates, feedback through the software, and cell phone alerts.

✔ **Objective 5.4: Triggering Devices**   Triggering devices such as light break sensors, motion sensors, and other types of sensors and triggers are used to start and end the interrogation to achieve efficiency of the system and reduce possible interference.

✔ **Objective 5.5: Real Time Location Systems**   RTLSs are employed to track high-value or mission-critical objects. RTLSs employ an active tag that will assist a back-end system in calculating its position relative to known points in the supply chain. Types of RTLSs are DTOA (Differential Time of Arrival), RSSI (Received Signal Strength Indicator), time-of-flight ranging system, and amplitude triangulation system.

REVIEW QUESTIONS

1. Which of the following is a required characteristic of an EPC UHF Gen 2 printer?

   A. EPC Gen 2 tags do not require a printer, because they are preprogrammed.

   B. All Gen 2 RFID printers also contain an UHF antenna that implements Gen 2 anti-collision algorithm.

   C. All Gen 2 RFID printers also contain an HF interrogator because the tags are very close to the antenna.

   D. All Gen 2 RFID printers also contain an integrated RFID interrogator.

2. RFID-based RTLS systems are typically used for which of the following?

   A. Locating items in large areas such as yards

   B. Tracking rail cars across the country

   C. Locating airplanes in real time

   D. Locating lost luggage in real time by airlines

3. Automated encode, print, and apply label applicators can do what?

   A. Encode multiple labels at a time

   B. Encode more than 250 tags per second

   C. Only use RFID labels

   D. Precisely place labels on the cases

4. When the RFID printer prints half of the image across the consecutive labels, what is the first thing you should check?

  A. Label size setting

  B. RFID protocol setting

  C. Printer ribbon alignment with a label

  D. Label feed rate

5. Why is it important to know where the RFID inlay is located within a label? (Select two answers.)

  A. Timing from the leading edge is critical to the encoding process.

  B. Printing over the chip can damage it.

  C. The location of the inlay identifies the label orientation.

  D. The location of the inlay dictates the speed of the printer.

6. Which of the following printing methods are used by RFID printers? (Select two answers.)

  A. Direct thermal

  B. Laser jet

  C. Inkjet

  D. Thermal transfer

7. What is the definition of Differential Time of Arrival (DTOA)?

  A. An algorithm used in passive RFID tags to avoid collision

  B. An algorithm used in the interrogator to isolate tags arriving late in the session

  C. An algorithm used in real time location systems (RTLS)

  D. An algorithm used for RFID printer calibration

8. An RFID printer is printing the labels on the release liner rather than the label itself. What is the possible cause of this?

  A. The labels are wound clockwise on the roll, but a counterclockwise wound roll is required.

  B. The label roll is improperly loaded in the printer.

  C. The label roll is defective. Labels are mounted on the wrong side of the release liner.

  D. The labels are wound counterclockwise on the roll, but a clockwise wound roll is required.

9. In a distribution center, 20 dock doors have passive UHF RFID portals. Most dock doors are used infrequently, so you would like to limit polling for tags only if tags are in the IZ. Which of the following is the most reliable way to control scanning at the dock doors?

    A. Install switches at all dock doors to turn the interrogator on and off.

    B. Provide the forklift operators with remote control units to turn the interrogators on and off.

    C. Install software that turns scanning on and off based on operator commands.

    D. Install light break sensors to detect the presence of a forklift.

10. Before performing any maintenance on a smart label printer applicator, you must do the following:

    A. Disconnect the air supply.

    B. Make sure that the label applicator is grounded.

    C. Move the label applicator to the maintenance area.

    D. Let the label applicator cool down.

REVIEW ANSWERS

1. **D**   This question tries to confuse you. All RFID printers including Gen 2 RFID printers contain an integrated RFID interrogator. The Gen 2 protocol refers to UHF, as you learned in previous chapters; therefore, a UHF printer cannot use HF interrogator. RFID printers do not use anti-collision because they communicate only with one tag at a time so a collision (when two or more tags respond at the same time) cannot occur. Finally, EPC Gen 2 tags may or may not require a printer, depending on whether they are already programmed by a reader or whether they need to have information printed on the label.

2. **A**   RFID-based RTLS systems are typically used for locating items in large areas such as yards, where you can install the network of RTLS access points (readers).

3. **D**   Automated encode, print, and apply label applicators can precisely place labels on the cases. They cannot encode multiple labels at a time or encode more than 250 tags per second. Some devices can encode

around 250 tags per minute. These devices can also use non-RFID labels, where the labels will not be encoded but only printed and applied.

4. **A** When the RFID printer prints half of the image across the consecutive labels, the first thing you should check is the label size setting.

5. **A** **B** It is important to know where the RFID inlay is located within a label because timing from the leading edge is critical to the encoding process and printing over the chip could potentially cause damage to it.

6. **A** **D** Direct thermal and thermal transfer methods are used by RFID printers.

7. **C** Differential Time of Arrival (DTOA) is an algorithm used in real time location systems (RTLS).

8. **B** When an RFID printer is printing the labels on the release liner rather than the label itself, the label roll is improperly loaded in the printer.

9. **D** In a distribution center, 20 dock doors have passive UHF RFID portals. Most dock doors are used infrequently, so if you would like to limit polling for tags only if tags are in the IZ, you should install light break sensors to detect the presence of a forklift.

10. **A** Before performing any maintenance on a smart label printer applicator, you must disconnect the air supply.

Design Selection

	NEWBIE	SOME EXPERIENCE	EXPERT
ETA	4 hours	2 hours	45 minutes

After the solution concept has been approved, design selection is performed, including the consideration of hardware and software targeted for actual deployment. Design selection is a critical part of any RFID implementation, because without the appropriate design, hardware, and software, the implemented RFID system will not function properly. The steps of design selection and considerations discussed in this chapter adhere to standards and regulations as well as compliance with trading partner mandates, and selection of operating frequency, tags, interrogators, antennas, peripherals, and software.

Objective 6.1

Standards, Mandates, and Regulatory Compliance

When selecting hardware and software for your RFID system, you must follow international, regional, and local standards and regulations. These are often implemented to prevent device interference or to protect the general public. Standards and regulations are discussed in detail in Chapter 10.

In addition, you must follow any trading partner–mandated specifications if the system you are about to design is targeted to comply with a particular mandate. A mandate will often define a particular tag data format and the frequency of operation.

Travel Advisory

Mandates are not the only reasons why you should look at the system your partners are using. If your operation is not a closed-loop system or it does not specify requirements for customizing your RFID solution, you should aim for interoperability and compatibility with partners' and vendors' products throughout the supply chain to achieve the full benefit of your RFID system.

Why Comply?

Reason number *one* is that noncompliance will cost you money. New designs must be compliant with all regulations and standards applicable to the system and region of operation. Any noncompliance with regulatory bodies can lead to steep fines and interruption of business, not to mention a potential tarnishing of your reputation. It is always good business practice to ensure that the manufacturers of the hardware you purchase follow the standards applicable to your system. Properly certified and standardized equipment will make your life with regulations a lot easier. Reason number *two* is to help keep interference between your system and your neighbor's installed system to a minimum.

Objective 6.2 Frequency Selection

If a mandate or particular standard has not defined what operating frequency your system must use, you must decide which frequency is the best choice for your system and the results you want to achieve. When making this selection, keep in mind the size of the interrogation zone in which the tags will be read as well as the item on which the tag will be placed. Different frequencies have different read ranges and behaviors when using various materials.

High-frequency (HF) and low-frequency (LF) technologies are inductively coupled technologies that function in the near field. This limits the operational range for passive systems of this type to 24 inches (0.75 meters) or less for HF and 36 inches (1 meter) or less for LF. These are typical ranges for standard systems, so variations in this performance can occur based on the size/type of the antenna used and power provided. A large antenna used with a passive LF system will produce a useable range of about 6 feet (2 meters). Passive inductive systems utilize the electromagnetic portion of the wavelength and therefore perform better around metals and liquids compared to capacitive systems (ultra high frequency [UHF] and microwave).

UHF and microwave technologies are capacitively coupled; they use the electric field generated from the reader antenna and are designed to operate in the far field. Capacitively coupled systems can achieve a much greater useable range than their inductive counterparts can. Typical ranges for the operation of these systems are 6 to 12 feet (2 to 4 meters), sometimes up to 25 feet (8 meters). The read ranges and performance of various frequencies are discussed in detail in Chapters 2 and 3.

Let's look at a few specific examples of considerations you'll need to make. If you were reading tags on a conveyor line that is 36 inches wide, for example, you could successfully use HF technology because it has a maximum practical read range of 24 inches. This fact allows you to place an antenna on either side of the conveyor, covering the entire volume of space across that conveyor. In a different example, you may be installing a dock door portal knowing that the distance across the dock door is 10 feet (approximately 3 meters). You will choose either a UHF or microwave frequency system. Both of these systems can achieve the adequate distance to cover that volume of space, while LF and HF systems cannot.

The next phase of design selection is hardware selection. You must select the specific tags, readers, antennas, and peripheral devices appropriate for your installation. All these items must interoperate successfully to deliver a solution that will drive tangible business value for your client.

 **Objective 6.3** **Tag Selection**

Y ou need to evaluate many parameters when selecting tags for your implementation, such as tag type, read range, operational environment, data transfer rate, form factor, memory, alternative data storage in case of defective/ damaged tags, and what type of product will be tagged.

Read Range

As you learned in Chapter 3, tags come in three basic types: active, passive, and semi-passive. These tag types differ mainly by their read range and cost. Active tags have long read ranges and high prices compared to passive tags. Semi-passive tags fall somewhere in between, depending on their design and capabilities. Therefore, if you were trying to read over a great distance beyond a passive tag's capabilities and you could not justify the cost of an active tag that would offer even greater distance, you should select the semi-passive technology.

Active tags are typically used in real-time locating system and can read from hundreds of feet away. Active tags are expensive relative to passive tags. Passive tags are used when a low-cost, high-volume solution is needed to track items at relatively short distances. Semi-passive tags are used when you need a slightly longer read range and perhaps sensors that will record temperature, vibration, shock, humidity, and other environmental parameters.

Significant read range differences exist among the different frequencies used for active tags. Generally, an active tag's read range increases with decreasing frequency used (if the interrogator- and tag-transmitted power stays the same). The longer the wavelength, the farther the wave will propagate depending on available power. For instance, a tag operating at 433 MHz will achieve a farther read than a microwave tag using the same power output. The data transfer speed capabilities of various frequencies apply to active tags as they do to passive tags.

Travel Assistance

Recent advancements include UHF near-field technologies, which are changing much of the accepted UHF tag capabilities. UHF near-field tags provide read ranges from 1 to 3 feet and good performance around metals and water with high data transfer rates. This text was written in support of the RFID+ Certification Exam, which does not include data pertaining to this technology since it was not available at the time the exam was written.

At this point in the process, a frequency should have already been selected for the particular application you are designing. Given the varying read ranges of tags with different technologies and different frequency ranges, you'll find a large variety of tags on the market. Your choice will dramatically narrow once the frequency selection process has occurred.

Operational Environment

Knowing the conditions to which the tagged products or objects will be exposed as they are being interrogated can dramatically assist you in making decisions regarding the tags as well. If you are choosing tags for use in outdoor applications, for example, a rugged tag that can be exposed to temperature extremes, direct sunlight, and rain should be chosen. If your operation moves goods in and out of a freezer during the handling process, then the tags must be able to withstand wide temperature variation and possibly humidity. Tag manufacturers will provide datasheets for their tags, which will specify all of the environmental conditions in which the tag can operate.

In specific applications, the only way you'll be able to find the right tag will be to test it in the appropriate scenario. Consider sterilization, for example—a process used in both produce and pharmaceutical markets. Some types of sterilization will destroy standard tags, and tag manufacturers may not know whether your process (such as irradiation, high-pressure washing, or high-temperature steam cleaning) will affect a product or not. In such a scenario, you should ask the tag manufacturer for sample tags that you can run through the process in question to validate the tags' endurance and survivability.

Data Transfer Rate

The tag's data transfer rate requirement is driven by how much time the tag can spend in a particular read zone or how much data will need to be read from or written to the tag. The data transfer rate increases as the frequency of the system increases. Low frequency has the slowest data transfer rate; microwave has the highest data transfer rate.

In applications where high speed is not required, selection is more about penetration and range than performance. For example, animal tracking is typically done with LF technology, since it functions well around liquids (that is, large bodies such as cattle, zoo animals, pets, and so on) and are read as the animal is contained within a given area or walking past a reader. Another application of this technology is the use of RFID tags in ignition circuits. In this scenario, read speed is not critical because the key is in the car, close to the reader as it sits in the "On" position. If the tag is seen, the car starts. If it is not, the car assumes a counterfeit key is being used and it will not start.

In toll-tracking applications, the tags (usually UHF and microwave for their fast data transfer rates) usually transmit a simple identity, which is then processed by the back-end system to charge the user of this tag for traveling on a road or bridge. For toll systems, the active technology is often used for enhanced range so that the data can be processed before the vehicle passes through the portal constructed on the highway. Some toll systems use specialized passive UHF tags, such as the toll system in the Dallas/Fort Worth area of the United States. The system has to capture the tag ID. If the tag is not present or is expired, a camera system is triggered to take a picture of the vehicle and driver. To fulfill all of these functions in a matter of seconds, the RFID systems must have sufficient read range, fast data transfer rate, and very fast interface with the back-end system and a camera.

Form Factor

Many forms of tags are available on the market. You should base your choice on the tagging volume and application technology as well as the necessary protection from the environment and physical wear and tear. If you intend to tag boxes in an automated manner, for example, the tags should come embedded in labels for integration into existing printing and application systems. The same would apply to any high-volume tagging application, in which you need to attach a tag to an item or a shipping container quickly. If you were deploying a system at a movie theater or sports arena, for example, you would use tags that can be embedded within tickets that will be handed out at admission stands. Another popular form is a button tag (Figure 6-1) that can be sewn into laundry, attached to bags of money, or placed into cargo containers. For bottles of pharmaceutical products, a small tag could fit nicely on the side or bottom of the bottle, as opposed to tags of larger size that are designed for use on cargo containers and that must be rugged and readable from greater distances. Alternatively, you may require tags that can be embedded into pallets or plastic objects as they are being manufactured and can be used to track those same objects through the supply chain as well for warranty purposes when they return.

FIGURE 6.1   Button tag (graphic courtesy of Texas Instruments)

Memory

Memory is also a prime consideration during the design selection phase. You will choose a tag that provides an appropriate amount of memory depending on how much data and what type of data a customer needs to write to or read from the tag. Active tags and semi-passive tags can contain larger amounts of memory than passive tags, because these tags can keep a memory bank powered even when it is not being queried over a long period of time.

The industry standard for identification number size today is 96 bits. Most of the tags that house this 96-bit data identifier have a general memory capacity of 256 bits. Currently, passive tags available for use in the global supply chain have up to 2048 Kbits of storage available. Semi-passive and active tags usually measure their memory capacity in megabytes of data due to the on-board power. The source of on-board power may not always be a battery; for instance, it can be powered from a power source to which a vehicle's tag is attached.

Larger memory is also required by sensors that are incorporated into a tag to keep track of sensed conditions. This data can be either transmitted to an interrogator and from there to back-end systems in real time or kept in a tag's memory and transferred in batches on demand.

Most tags store a simple number that can be transmitted very quickly. This number is then looked up in a back-end data processing system for the data that correlates to that specific tag and item. In rare scenarios in which a back-end data connection is not feasible, you may find tags that store more data directly on the tag itself. Examples of this can be seen in medical applications where HF semi-passive tags are used to measure the temperature of blood bags. The tags can store upwards of 600 data points, and they can provide this data on demand without any back-end data connection.

Type of Product

When selecting the tag you will be using in your implementation, you must consider the type of product that will be tagged. Some tags are specifically designed for products containing metal, others are specifically designed for products containing high moisture materials, and some generic tags are built for application onto corrugate that is used for packaging.

Passive tags' operation depends on efficient use of power. The response from a tag back to the reader can be from a millionth to a billionth of the original power transmitted to it. This makes proper tagging critical to the reading success. All tags undergo a certain amount of detuning when applied to items. For the most optimal use of power, the resistance of a tag's antenna should match the resistance across the chip attached to it. When these values differ, the tag is wasting some power due to the mismatch.

Local Lingo

Detuning   *Detuning* is a change in the resistance across the antenna, which will change the efficiency of the tag.

Tags designed for use on metal (as well as other specialized tags intended for use on glass, plastics, wood, and other materials) will have a calculated mismatch in the resistance of the chip and antenna (see Figure 6-2). When the tag is applied in close proximity to metal, the detuning that occurs will bring these numbers much closer together. However, this tag cannot be used on nonmetallic surfaces because it would not get detuned and therefore would not perform according to specifications. The same considerations apply to any other specialized tags.

FIGURE 6.2   Tag designed for use on metal (graphic courtesy of Intermec)

Region

As the global RFID market emerges, it will be necessary for tags to be manufactured that will respond reliably in all regions of the world within the UHF band from 860 to 960 MHz. In the past, tags were designed to resonate at a specific frequency, usually the center of the targeted band. Now manufacturers are working to create tags with a center band with a very wide range.

Figures 6-3 and 6-4 show network analyzer screens that are measuring response power at four markers spread across 860–960 MHz. In Figure 6-3, the first marker is at 867 MHz for European markets, the second and third markers are 902 and 928 MHz, respectively, to mark the beginning and end of the North American band, and the fourth marker at 956 MHz represents the Asian market band. Each horizontal line represents a 5 dBm variance from the line above or below.

Travel Advisory
Remember, dBm measures changes in power from a known source. A 3 dBm increase equates to double power, and a 3 dBm loss equates to half power.

As you can see in Figure 6-3, the tag represented would perform best in the Asian market, not very well in the European market, and it would have very sporadic results in the North American market. A 5 dBm difference in power from the start of the North American range to the end of the North American range would show a dramatic difference in performance as the reader hops between frequencies in this range. Your goal is to provide reliable, predicable results as your product ships throughout the world. Figure 6-4 presents a tag that provides relatively equal performance around the world (a *world tag*). (The testing shown here, as well as the photos, are provided compliments of Venture Research in Plano, Texas.)

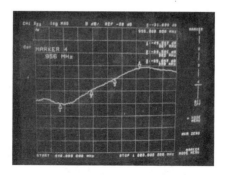

FIGURE 6.3   Uneven power response across 860–960 MHz

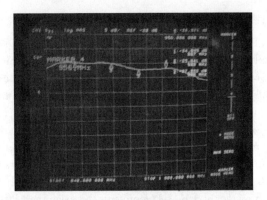

FIGURE 6.4   Fairly even power response across 860–960 MHz

Local Lingo

World tag   A *world tag* offers little variance between the frequencies recorded, providing consistent results irrespective of where in the world it is used. Therefore, it is suitable for tagging objects moving through an international supply chain.

Objective 6.4 Interrogator Selection

After choosing the appropriate tag for your installation, you must ensure that the interrogator you select can operate on the appropriate frequency, can support the protocols of the tag, has the capability to integrate with the back-end system through the appropriate interfaces, and can survive the conditions associated with the environment within which it will operate.

The frequency of the interrogator should have already been chosen when you selected an operational frequency for your tags. The frequency of the interrogator must match the frequency of the tag or your system will not work.

The protocols supported by the interrogator must all match the tags you selected for your implementation. Likewise, if you are accepting tags from trading partners, your interrogators must also support the protocols that your trading partners use. Several multi-protocol readers on the market will allow you to choose one or more protocols, according to your needs. However, if you choose more than one protocol, the reader will perform slower because it has to run through all chosen protocols when scanning for the tags in the interrogation zone. In addition, it is a good idea to ensure that your readers can upgrade or change protocols (usually through firmware upgrades) based on changes in the market, so that you do not end up with an obsolete reader as new protocols emerge.

Readers can be also selected according to their communication capabilities. Many readers today have general-purpose I/O (GPIO) ports that can be configured to drive specific applications and that have direct integration capabilities with light stacks, light break sensors, and a wide variety of I/O devices. Also, many readers have internal processors and operating systems that can run software directly on the reader, which lets it act as an edge server that will filter and aggregate data prior to handing it off to the back-end system.

The interfaces the reader supports must match the network that you have installed currently at a facility. If you have a wireless network employed and want to add your readers to it, they must support the security methods used on your network. If the readers do not have wireless capability built in, external wireless adapters can be used that connect to any standard Ethernet port, which are present in most readers.

Another important consideration is the environment in which the reader is being installed. If the reader is being mounted on a forklift or on a material-handling vehicle, for instance, it will be exposed to shock and vibration, and most consumer electronics are built to withstand only 5Gs of shock. The amount of shock experienced by a fully loaded forklift as it drives around the warehouse can be up to 20Gs on a regular basis. If the RFID equipment will be exposed to these types of conditions, they must be certified by the manufacturer to survive if you expect them to live in that environment for very long. To see what happens when the environmental factors are not properly considered, see Figure 6-5.

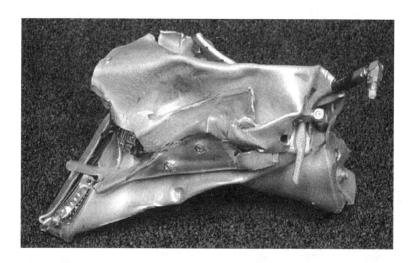

FIGURE 6.5 Remains of a reader when environmental factors are too intense

Objective 6.5   Antenna Selection

After choosing the appropriate reader for your application, you must select the antennas. Antennas are selected based on their polarization, gain, ruggedness, size, cable, and mounting options.

Polarization

Antennas can be either *circularly* or *linearly* polarized. Circular antennas are used in implementations for which the orientation of a tag cannot be assured. Circular antennas typically have a shorter range than their linear counterparts; however, they will assure that the tag is readable regardless of the angle/orientation of the product/container at which the antenna is directed.

Linear antennas read larger distances than circular antennas but require tagged objects to be in the same orientation every time they are presented to the antenna. A good example of a linear antenna's use would be on a manufacturing line, where machines are placing objects on conveyors for movement through a facility. If a machine places any object on a conveyor line, each object should be placed in the same orientation and position on the line. Linear antennas offer better penetration of dense objects—consider, for example, a company reading tags through rolls of paper. Bulk paper is dense and contains a lot of moisture; therefore, a linear antenna stands a much better chance of penetrating the whole role than a circular antenna.

Exam Tip

Know that to achieve even better results, you can place a right-hand circular polarized antenna next to a left-hand circular polarized antenna. Assuming they are run by one reader, the waves will always be at different angles and should provide a more efficient read performance.

The polarization of an antenna is discussed in detail in Chapters 1, 2, and 4.

Gain

The gain of an antenna plays a huge role in the results your interrogation zone will achieve. The higher the gain of an antenna, the narrower the beam of that same antenna. A higher gain means that you will be able to read tags at a greater distance with this antenna. However, higher gain also means that the antenna will provide a narrower beam; therefore, if the tags are in motion while being read, they will spend less time in a narrower beam than they will in a wider

FIGURE 6.6 Low-gain antenna beam

beam. If you need to read tags at relatively short distance but want to achieve a longer dwell time, you should use an antenna with a lower gain, because this will cause the beam to be wider but with a relatively short read range. Typical supply chain antennas for passive RFID systems have a gain of 6 to 8 dB. (Antenna gain is covered in detail in Chapter 2.) Low- and high-gain antenna beams are represented in Figures 6-6 and 6-7.

HF and LF antennas are not directional. They propagate a signal equally in a manner that is perpendicular to the coil of the antenna. To increase the range of the antenna, you must increase the size of the coil or the power provided to it. The read range of these frequencies is driven by the practical size for an antenna. Installation and maintenance of huge antennas is more costly and more problematic than simply using a frequency more suited for a specific application.

Ruggedness

An antenna must be selected based on its abilities to withstand its environment. Lightweight and inexpensive antennas can be used in retail environments that are temperature controlled and have protection built around the antenna. For example, the antennas can be built into a countertop for a cash register, where they will never be touched by the goods while tags are being read.

However, if you are mounting antennas at a warehouse dock door, they are often exposed to sun, rain, snow, wind, and other elements. The antennas are also exposed to vehicles driving by and personnel working nearby, all of which can accidentally run into them. Antennas for use on a forklift need to be extra rugged and

FIGURE 6.7 High-gain antenna beam

able to withstand direct impact from cargo on the forks. A rugged antenna may be needed on baggage handling devices that can be used indoors or outdoors. For such conditions, you must select a rugged antenna if you expect it to survive without additional protection. Covers can be also used to protect exposed antennas.

Size

The size of the antenna must also be considered. The size must be practical and fit the application you are designing. If the antenna is to be embedded within an object, obviously the antenna must be smaller than that object.

Although bistatic readers offer better performance in difficult read situations, their use may be impractical due to a relatively large size of a bistatic antenna compared to a monostatic antenna. A bistatic reader needs two antennas (in separate cases or joined in one case) to function, while the monostatic reader will utilize one smaller antenna.

Mounting Options

Mounting options available with the antenna could also help you with your selection. Check out what the specific antenna models offer (holes for screws, flexible joints, and so on) to determine which mounting option will be the best for your system. You may want to use screws, zip ties, adhesives, industrial Velcro, or other options. Breakaway antenna mounts are similar to the breakaway mirrors installed on cars—these are certainly advisable if you know that the antennas may be occasionally hit by a vehicle or other equipment. Mounting options will not change your antenna's performance but will be important for installation and long-term maintenance.

Antennas can be mounted in readymade dock stands that hold from two to four antennas (and a reader and certain I/Os), which are protected by an impact-resistant cover. Antenna mounts designed for forklifts embed antennas near the mast of the truck and offer a good degree of protection as well, ensuring antennas are close to the goods the truck is carrying and yet protected enough to expect a reasonable lifespan. For antennas on conveyors, extruded aluminum frames can be constructed that surround a section of conveyor to allow antennas to be safely placed in close proximity.

Travel Advisory

You can mount this equipment in your facility in many ways. The factors that will drive your decision are cost, form factor, and environmental survival. Note that if you let cost be the primary factor, and you go for low-cost but inappropriate solutions, your selection may cost you a lot more over the long term.

Objective 6.6 # Cable Selection

Typically, RFID antennas come with cables enclosed. These cables can be used when the antennas are within the distance of the cable length from the interrogator. However, in extreme circumstances, where the antenna may be placed well beyond the distance of the original cable, additional cable can be purchased and inserted into the path from the reader to the antenna. The cables used to feed RFID antennas are coaxial cables with a center conductor that is solid wire and a multi-stranded, braided outer conductor that is separated by a dielectric.

If you are sending a signal across a great distance, a thicker cable is more advantageous as it loses less of the signal. Although all cables experience loss, thinner cables have a higher amount of loss than thicker cables at the same frequency. The disadvantage of thicker cables is that they have a smaller bend radius than thinner cables and thus cannot be used in applications where the conduit must bend 90 degrees or more.

Exam Tip

Remember that the thicker and shorter the cable, the lower the signal loss and the lower the bend radius.

 Objective 6.7 # Peripherals Selection

You can add peripherals to an RFID system to enhance its capabilities, such as RFID-enabled printers, smart label printer applicators, light stacks, triggering devices, and extrasensory networks. The integration of a well-planned system can include many different types of input and output devices.

RFID Printers and Label Applicators

Printers are used for rapid integration into existing systems where barcodes may be used. An RFID-enabled printer can easily replace an existing label printer, providing the production line with the ability to place RFID tags onto objects they regularly handle. Smart label printer applicators can put labels on products in an automated fashion if the products are all of the same approximate size and presented at a regular interval. An automatic smart label printer applicator can certainly increase the efficiency of RFID tagging. However, for lower tagging volumes, you may select a handheld label applicator or manual application.

Printer variables such as protocols, interfaces, label width, error handling, and others are discussed in detail in Chapter 5.

Light Stacks and Triggering Devices

Light stacks, horns, and display devices are used for notifying an operator about a particular condition the system may be experiencing. Choosing the appropriate alert will ensure the correct action is taken to correspond to the condition. Choosing the wrong alert will result in missed events and an ineffective system.

If you are sending an alert, ensure that it is seen or heard. In brightly lit areas, an audible alert may be best. In noisy environments, a visible alert should do the job. Because of variable conditions, no absolute correct selection criteria exist. You should, however, ensure that the signal is not wasted and that the notification devices can survive the environmental conditions in which they are placed.

Triggering devices are used to let the system know when a tagged object has entered the area. Triggering devices can be proximity sensors, motion detectors, light break sensors, weight sensors, button presses, and other devices.

Objective 6.8   Software Selection

The software used to run and integrate the RFID system must be compatible with the hardware that you selected and the back-end systems to which you are integrating. Standards for RFID middleware and data management software are still emerging; therefore, when selecting software today, ensure that it will be upgradeable in the future and that your in-house staff can support it.

In an enterprise-wide deployment, hundreds of devices will have to be managed by a single middleware platform. Therefore, when choosing middleware, you must ensure that it will be able to handle the workload. Some of the other features you may want to look into are device health status monitoring (for instance, using Simple Network Management Protocol (SNMP)) device management capabilities, data filtering and aggregation, and automated exception handling. Software licensing and maintenance costs in a large installation can be very substantial and must be considered early on in the planning stage.

Objective 6.9   Shielding and Equipment Protection

Once you have a more complete picture of the overall design, you can look at the overall environment in which the system is operating upon installation.

Electrical noise and interference, vibration, temperature and humidity, indoor or outdoor installation, and potential system abuse are all key points of consideration for the environment.

If your interrogation zone is near a conveyor with large motors or a number of other readers are nearby, it may encounter RF interference. To solve this problem, you may need to install shielding material such as chain-link fence or anechoic foam (Figure 6-8). Grounded chain-link fence makes an excellent RF shield as long as the distance between intersections is less than a quarter the wavelength of the frequency being used. In addition to shielding, you can install software to manage and synchronize the readers and other equipment to reduce or eliminate interference.

The vibration, temperature, and humidity to which the equipment will be exposed will help you determine the need for rugged equipment, reader enclosures (such as Ingress Protection (IP) or NEMA—National Electrical Manufacturers Association), or possibly special enclosures with ventilation or temperature controls built into the box.

If the equipment is being installed where it is exposed to outdoor conditions, you should take extra precautions to protect the equipment from environmental elements. In indoor installations, the equipment can suffer from electrostatic discharge (ESD), especially in extremely dry environments. For environments that are excessively prone to ESD, you can ensure that equipment and tags are not damaged:

- Tags should be stored in antistatic (conductive) bags (Figure 6-9).
- Areas in which a lot of sensitive electronics operate should have conductive floor mats installed.
- Particularly problematic areas can be treated by installing air ionizers to reduce the probability of static discharge (Figure 6-10).
- All electronics that are installed in a complete system should be grounded to minimize the potential for static damage.

FIGURE 6.8 Anechoic foam

FIGURE 6.9   Antistatic bag

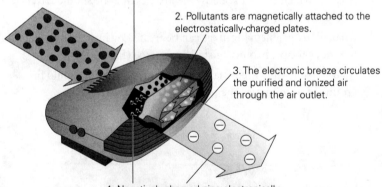

1. The electronically-created breeze draws pollutants into the air inlet.

2. Pollutants are magnetically attached to the electrostatically-charged plates.

3. The electronic breeze circulates the purified and ionized air through the air outlet.

4. Negatively-charged pins electronically circulate trillions of air-cleaning negative ions throughout the room.

FIGURE 6.10   Ionizer

CHECKPOINT

✔ **Objective 6.1: Standards, Mandates, and Regulatory Compliance**   When selecting hardware and software for your RFID system, you must follow international, regional, and local standards and regulations to ensure proper performance, interoperability, and safety. You may also need to follow mandates supplied by your trading partners or customers.

✔ **Objective 6.2: Frequency Selection**

	Passive				Active	Semi-passive
	LF	HF	UHF	Microwave	All	All
Read/Write Distance	2–3 ft.	1–2 ft.	15–20 ft.	5–10 ft.	Up to 1000 ft.	Up to 300 ft.
Tag Response Time	Slow	Slow	Medium	Medium	Fast	Fast
Data Transfer Rate	Slow	Medium	High	Highest	Depends on frequency	Depends on frequency
Storage Capacity	Low	Low	Low	Low	High	High

✔ **Objective 6.3: Tag Selection**   Tag selection is so closely tied to frequency selection that the Objective 6.2, "Frequency Selection" table works for tag selection too! Make a point to know this table well.

✔ **Objective 6.4: Interrogator Selection**   You must select the interrogators according to environment, mandates, and standards and their performance will depend on the frequency, whether the system is active or passive, what power is supplied, the type of antenna and tag used, functions of the interrogator (filtering, anti-collision, dense reader mode), as well as interference from other interrogators or the environment.

✔ **Objective 6.5: Antenna Selection**   Antennas are selected based on their polarization, gain, ruggedness, size, cable, and mounting options.

- Linearly polarized antennas are better when the tag orientation is known, and they penetrate dense objects better and achieve longer read range when compared to a circular antenna with the same power input.
- Circularly polarized antennas are better when the tag orientation is not fixed.

- Higher gain antennas are better for achieving longer read ranges with a more focused beam, while lower gain antennas are better for achieving a wider beam and therefore a longer dwell time.
- LF and HF antennas are not directional.
- You must select antennas according to the environment in which they will be used—on vehicles or by dock doors, use rugged antennas or special protective measures such as covers.
- When concerned about size, choose monostatic antennas rather than bistatic.

✔**Objective 6.6: Cable Selection**   When selecting a cable, remember that the longer the cable, the greater the signal loss. The thicker the cable, the lower the loss but the lower the bend radius. The higher the frequency, the higher the loss. Therefore, you should use shorter cables if possible, but if you must shorten them more than the manufacturer recommends, you must lower the interrogator power output so as not to surpass the limit for radiated power.

✔**Objective 6.7: Peripherals Selection**   To enhance your RFID system, you can use the following peripherals: RFID-enabled printers; smart label printer applicators for printing; smart label encoding and application; light stacks, horns, and other feedback devices to signal changes in the system performance; triggering devices to start and stop interrogation; and extrasensory networks to provide the RFID system with the ability to monitor movement, temperature, and other environmental conditions.

✔**Objective 6.8: Software Selection**   When selecting software for your RFID systems, look for software with device health status monitoring, device management capabilities, data filtering and aggregation, automated exception handling, and other similar features.

✔**Objective 6.9: Shielding and Equipment Protection**   Equipment must be protected from physical damage as well as interference. Interference can be caused by other readers in the environment but also any other devices that are working on the same or similar frequency, such as wireless LANs, large motors, microwave ovens, cordless phones, and cell phones or malfunctioning devices that create the same frequency signal, such as speakers, conveyors, forklifts, and so on. To solve interference problems, you may have to install shielding material such as a chain-link fence or anechoic foam. In addition to shielding, you can install software to manage and synchronize the readers and other equipment to reduce or eliminate interference.

For antennas on conveyors, use extruded aluminum frames; antennas at dock doors should be encased into rugged stands with transparent covers in front of the antenna. Antenna mounts designed for forklifts embed antennas near the mast of the truck, which offers a good degree of protection.

REVIEW QUESTIONS

1. When two circularly polarized antennas are facing each other and may operate at the same time, what can you do to reduce interference between them?

   A. Use one antenna with right-hand polarization and the other with left-hand polarization.

   B. Use both antennas with the same polarization.

   C. Increase power at one of the antennas.

   D. Replace the second antenna with a linear antenna.

2. You are asked to implement UHF RFID interrogation zones to account for pallets received and shipped from eight dock doors in a row adjacent to each other. The dock doors are 8 feet wide and separated from each other by 10 feet. Each dock door needs only one antenna at the top of the door to read tags. The tags will always be at least 4.5 feet from the antenna while they pass through the dock door. What is the most effective solution using a minimum number of readers, computers, and cable?

   A. Use eight readers and eight computers, one per dock door. Connect each reader to a computer using a serial cable. Connect all computers to a LAN.

   B. Use two readers, each with four antenna ports. Install one antenna per door and run the antenna cable to a reader. Connect two readers to a computer and the computer to a LAN.

   C. Use four readers, one per two adjacent dock doors, with at least two antenna ports and an Ethernet port. Connect two antennas to a common reader, and connect all readers and the computer into a LAN.

   D. Use eight readers with an Ethernet port and connect them to one computer, and connect the computer to a LAN.

3. You are asked to assist with the design of an RFID system that will be used in a large facility to track personnel, with a resolution of 20 feet. Which of the following systems would you recommend?

   A. Active real time location system (RTLS)

   B. Passive HF RFID system with interrogators located at entry points

   C. GPS

   D. Passive RFID system with personnel scanning using handheld interrogators

4. Which of the following RFID systems would provide the largest read range when tags are attached to a metal freight container that is 40 feet long?

   A. A passive 915 MHz tag and interrogator operating at 1 watt
   B. A passive 13.56 MHz tag and interrogator operating at 1 watt
   C. An active 433 MHz tag and interrogator operating at 20 milliwatts
   D. An active 2.4 GHz tag and interrogator operating at 20 milliwatts

5. Which of the following is an advantage of an interrogator's ability to multiplex its antennas?

   A. Angling the antennas to reach hard-to-read areas
   B. Cycling of the antennas to prevent wear and tear
   C. Reading tags at a faster rate
   D. Reading multiple protocol tags in the interrogation zone

6. Which of the following is *not* a consideration when selecting interrogator antennas for an RFID system?

   A. Antenna gain
   B. Antenna protocol
   C. Antenna frequency
   D. Antenna polarization

7. In an RFID smart shelf application, HF is a better technology than UHF because:

   A. HF antennas are not directional.
   B. HF antennas are directional.
   C. HF antennas have a shorter read range.
   D. HF antennas are less expensive.

8. In UHF RFID installation, a linearly polarized patch antenna is replaced with a circularly polarized patch antenna with the same gain. What happens to the RF energy emitted?

   A. RF energy available to tags is increased by 3 dB.
   B. RF energy available to tags is reduced by 3 dB.
   C. RF energy available to tags is reduced by 2 dB.
   D. There is no change in RF energy available to tags.

9. You are designing a UHF RFID system for use at the receiving dock door for reading pallets and cases. The cases on the pallet are of

different sizes and their tag orientation varies from one box to another. What is the best type of antenna to use for this application?

A. Inductively polarized

B. Horizontally polarized

C. Circularly polarized

D. Linearly polarized

10. Which of the following statements about coaxial antenna cable is false?

A. Cable loss is inversely proportional to the cable diameter.

B. Cable loss is directly proportional to the frequency of transmission.

C. Cable loss is directly proportional to cable length.

D. Cable loss is inversely proportional to cable length.

REVIEW ANSWERS

1. **A**   When two circularly polarized antennas are facing each other and may operate at the same time, to reduce interference between them you should use one antenna with right-hand polarization and the other with left-hand polarization.

2. **C**   The most effective solution using a minimum number of readers, computers, and cable would be to use four readers, one per two adjacent dock doors, with at least two antenna ports and an Ethernet port. Connect two antennas to a common reader and connect all readers and the computer into a LAN.

3. **A**   To track personnel in a large facility with a resolution of 20 feet, you should use the active RTLS.

4. **C**   The active 433 MHz tag and interrogator operating at 20 milliwatts would provide the longest read range.

5. **A**   The advantage of an interrogator's ability to multiplex its antennas is the ability to angle the antennas to reach hard-to-read areas.

6. **B**   Antenna protocol is not a consideration when selecting interrogator antennas for an RFID system. The protocol is a consideration for interrogators.

7. **C**   In an RFID smart shelf application, HF is a better technology than UHF because HF antennas have a shorter read range than UHF.

8. **B**   When a linearly polarized patch antenna is replaced with a circularly polarized patch antenna with the same gain and input power, the RF energy available to tags is reduced by 3 dB.

9. **C**   The best type of antenna to use for this application will be circularly polarized. If the orientation was not changing, the linear antenna would be a possible choice as well.

10. **D**   Coaxial antenna cable loss is not inversely proportional to cable length but is directly proportional.

Installation

ETA	NEWBIE	SOME EXPERIENCE	EXPERT
	4 hours	2 hours	45 minutes

179

In this chapter, you will learn about installation techniques for successful RFID implementation. RFID installations should follow the same basic principles of other technology implementations, such as wireless networks. The big difference between RFID installations and those of other technologies is the configuration of the actual interrogation zones (IZs), since understanding IZs will be critical to your installations' success.

Objective 7.1   Preinstallation

Prior to installation, you should study your installation manual and a site diagram. You should also perform basic equipment testing as well as other necessary network checks to ensure a correct and successful installation.

Site Diagrams

Typically, an installer will receive a site diagram with the installation documents. This can range from a photocopy of a fire escape plan that has been marked up to show installation points to a full-blown blueprint that has been modified professionally to indicate power outlets; data connections; heating, ventilation, and air-conditioning (HVAC); wiring closets; safety walkways; and other important aspects. An example of a site diagram is shown on the next page.

Preinstall Equipment Testing

It is always a good idea to make certain that the equipment you will be installing in the field works prior to bringing it to the site. By taking this extra step, you minimize the onsite time needed for installation, therefore causing as little disruption on the site as possible. To test and prepare the equipment for installation, follow these steps:

1. Assemble, set up, and configure all peripherals and ancillary devices with the appropriate settings. This includes Programmable Logic Controllers (PLCs), input sensors, and output devices.

2. Configure all general-purpose I/O (GPIO) settings on the interrogator and connect the peripheral devices to the interrogator.

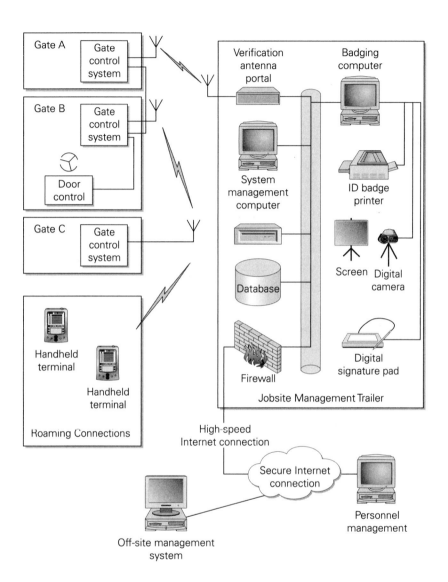

3. Install a local PC configured with the client's network data.

4. Connect the interrogator and PC to a local communications hub and send a ping or query to the device ensuring successful communication.

When setting up the peripherals prior to connecting them to an interrogator, you should always "breadboard" the components (connect them using short test cables on a tabletop without mounting) before mounting them into place and connecting to power. Sometimes, you may need the components to be placed in a National Electrical Manufacturers Associations (NEMA) enclosure. In this case, you should do the following:

1. Breadboard the components first to ensure that everything functions as expected.

2. Lay out the enclosure in a logical order to ensure that the equipment has the appropriate clearance around it to accommodate cable bend radius and that proper ventilation and maintenance access are available. This will also help you determine whether the cable lengths and types are appropriate for the installation.

3. Physically mount components to the backboard of the NEMA box and install the assembled backboard in the box.

4. Connect all components.

5. Connect to power.

After configuring the GPIO of an interrogator and connecting the peripherals, you should test the system by attempting to read a tag. This should power the appropriate output device. If this does not occur, you should do the following:

1. Check all connections.

2. Check configurations.

3. Test individual components.

4. Replace or reconfigure nonfunctioning elements and solve the problem!

When configuring communication settings in order to connect with the client's network, you will need to know and set up the following:

- For Ethernet or wireless Ethernet communications, configure the IP address, subnet mask, gateway, and appropriate security settings.

- For serial communications, configure the port settings such as the baud rate, bits per second, parity, stop bits, and flow control.

After you are done with the configuration, reboot the device and ensure that the settings were saved.

Once all of these steps have been completed, place all the equipment in the shipping crates and ship it to the installation site (be sure to package the components properly to prevent any damage in transit). By pretesting and preassembling the equipment, you ensure that the installation will be much easier for your clients and much easier on your installers, and the time you spend onsite will be more effectively utilized.

Preinstall Checks

Upon arrival at the installation site, a number of checks should be performed to ensure a smooth installation. You should determine the following:

- Whether you have the proper ladders, lifts, and safety equipment needed to reach the equipment that you are hanging overhead
- Whether all antenna locations or proposed locations for the interrogation zone have adequate access for optimization of the antennas as well as long-term maintenance
- Whether the antenna and its angle and height can or should be varied for the application
- Whether the antenna radiation can cause multipath reflections, for instance due to metallic structures or surfaces nearby that can reflect RF waves (these structures and surfaces may need to be covered with electromagnetic interference [EMI] absorbing materials)
- Whether readers can be placed in close proximity to the antennas they power (to avoid signal strength loss due to long cables)
- Whether the readers can be placed in such a way that the status indicators are easily visible to an operator for troubleshooting and long-term maintenance
- Whether the equipment is close to the travel path of vehicles or personnel (you may need to apply additional protective measures for the equipment)
- Whether appropriate cables and wiring are installed in the right place or adequate space exists to run new wiring as required
- Whether all cabling and wiring complies with industry standards

- Whether all brackets and mounts are in place, or you have space to install them
- Whether the power line is properly grounded

Travel Assistance

You must ensure that readers are mounted per the manufacturer's instructions and are well protected against physical damage.

As you do with any system installation, you have only a certain amount of time to get the job done. Therefore, to avoid interruptions and delays, you must ensure that you have all the tools you need to finish the job. For example, you may need screwdrivers, wire cutters, zip ties, a soldering iron and solder, black tape, and power tools such as drills or saws. Take a quick inventory of your equipment to ensure that you have everything necessary for the installation, as well as some spare equipment in case you need it. Don't forget spare RFID readers, connectors/adaptors, and cabling.

Exam Tip

A best practice is to make sure that the cable length from the reader to the antenna matches wave specifications for a particular frequency. Basically, you should cut the cable in increments of a half or a full wavelength of the frequency you are using.

Objective 7.2 # Hardware Installation

An RFID system can require the installation of antennas, RFID readers, RFID-enabled printer applicators, and RFID peripherals. This equipment can be also installed in larger configurations such as RFID portals or tunnels, as shown in Figure 7-1. These may be built into devices such as dock doors or personnel doors, conveyor systems, sortation devices, encoding stations, or forklift trucks.

FIGURE 7.1   Freestanding pallet portal

Antenna Installation

An antenna is one of the most critical components in an RFID system. It is also the most exposed part. Antennas must be close to the goods as they are being read. You need to know the antenna's orientation, approximate read range, and the footprint of the area that it will cover. Most antenna manufacturers include a datasheet or whitepaper that you can use to estimate the radiation pattern. If you are using multiple antennas, your antenna read zones should overlap slightly to enhance the probability of obtaining the reads as the tags travel through that IZ.

Mounting Antennas

The best practice for mounting antennas is to mount them on *a flexible joint* of some sort. By using a joint, you can adjust the antenna angles and modify your read zone area, and the antenna will give way with little pressure to swing away from potential harm, as goods or vehicles could impact it during their travel through the read zone.

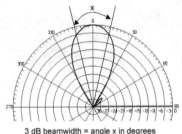

3 dB beamwidth = angle x in degrees

In the case of a monostatic reader, one antenna is used for both transmitting and receiving functions. In the case of a bistatic reader, two separate antennas and two separate ports perform these functions. If you use separate transmit and receive antennas, the corresponding pairs should not be mounted directly opposite each other. Because the transmit antenna transmits a powerful RF signal and the tags respond with a little power, the receive antenna would not be able to differentiate between the original transmission and the tag's response.

Travel Advisory

Do not disconnect an antenna when the read mode of the reader is turned on!

If the interrogator comes with terminators, be sure to place them on unused antenna ports before turning on the unit. It is a best practice to protect all antennas with RF transparent panels and covers, concrete bollards, breakaway mounting brackets, and other protective equipment. (We will mention this when discussing portal installation in the "RFID Tunnel Installation" section later in this chapter.)

To determine whether the radiation patterns from each read zone overlap, you should walk the space around your zone with a working tag to see if it can be read outside the IZ. If this is the case, adjustment of power or installation of RF-absorbent shielding materials may be required.

Antenna Cable Considerations

When installing antennas, you must pay special attention to cables that connect antennas to the reader. Feeder cable selection is important to limit reductions in antenna output. Cables will attenuate RF signals (the longer the cable the larger the attenuation or loss); therefore, keeping the cables as short as possible is important. If you require longer cables for your installation, you will need to select lower loss cables, which will increase power delivered to the antenna as well as the cost of the installation.

Attenuation values as well as the minimum bend radius for the cables are available from the cable manufacturer. The general rules for cable performance are as follows:

- The thicker the cable, the smaller the loss experienced during the travel of the signal on the cable.
- The thicker the cable, the smaller the minimum bend radius.
- As frequency increases on a cable, the amount of attenuation also increases. A 13.56 MHz system, for instance, will experience less loss on the exact same cable than a 915 MHz system.

Reader Installation

You should follow these rules when installing readers:

- Mount readers securely so that they are not disturbed, bumped, or damaged during regular business operations.
- Leave 5 to 6 inches of clearance on all sides of the reader, except on the bottom of a reader that has shock-absorbing mounts (the rubber feet attached to the bottom of most interrogators). This clearance is necessary in order to avoid cable strain when plugging cables into the reader, especially if you run them out of the box through conduit with the rest of the cables.

- Ensure that the reader is placed and installed in accordance with the recommendations of the manufacturer to keep your warranty valid.
- If environmental enclosures (such as IP or NEMA enclosures) are required for the protection of the readers, make sure that you use a correct type of enclosure and that you install it correctly.
- Place readers in a way that will make the status indicator lights or LEDs visible to an operator or other personnel; this can assist you with exception processing and troubleshooting. If you are using an enclosure, you may need to install a window or a light panel/stack for notification of a device failure.

Photograph courtesy of
Intermec Technologies

Travel Advisory

When installing electronics, you should couple them with a triggering device that will ensure the unit is active only when needed. It is best practice to turn off devices when they are not in use, as this will increase the lifespan of the equipment, lower ambient RF noise in the facility, and lower the mean time between failure (MTBF) of the equipment.

After Reader Installation

After installing a reader, you should perform a few simple checks to make sure that the reader functions correctly:

- Perform a simple power-on test to ensure that all the lights and appropriate feedback mechanisms are connected and talking to the reader.
- Make sure that the reader is connected to the host system correctly. You can do this typically by pinging the device across the network to ensure the network connectivity.

Local Lingo

Pinging *"Pinging* the device" refers to sending a ping command to a networked device using its IP address from a host in order to find out if it responds or not. If the networked device is communicating, the response to a ping command will typically be a reply with data packets. If the device is not communicating, the host will see "Request timed out".

- Validate the reader's configuration parameters. Make sure nothing has changed from the time you set it up and that you can successfully read tag data. This test is usually performed prior to connecting the reader to the back-end data network.

- If an extra antenna cable is used, wind the whole cable length into a coil to keep the cable out of harm's way. Feeder wires running from the reader to the antennas should be of approximately equal length to ensure equal power delivery to transmit and receive pairs of the antennas.

- Ensure that the cables are not running through the same conduit or in the immediate proximity of power lines; otherwise, interference in the system could occur along with an inconsistent delivery of power to the antennas.

Photograph courtesy of
Intermec Technologies

Ancillary Devices Installation

Once the readers and antennas have been installed, you can install the ancillary devices—photo sensors, pressure sensors, motion sensors, printers, label applicators, and other devices. Follow the general rules for antenna, reader, and printer installation (for printer installation, see Chapter 5) by following the device manufacturer's installation guidelines.

Interrogation Portal Installation

A portal can be used at the dock door, the personnel door, or any kind of doorway, entrance, or exit. A portal reader configuration is intended to identify objects with RFID tags as they pass through the doorway entering or leaving a facility. An example of an interrogation portal is shown in Figure 7-2.

Tags are often mounted on objects, cases, or pallets that are being moved in and out of a distribution center of a warehouse. Tags can be also mounted on material handling devices, tools and equipment, as well as people and other assets that are important to the business. The location or the last location of these objects can be tracked as they pass through the interrogation portals.

FIGURE 7.2   Interrogation portal

Note that the height of objects traveling through the portals can vary. For instance, if a distribution center were to ship double-stacked pallets, an additional set of antennas may need to be added to the top of the portal to cover the volume of space that the second pallet will occupy.

Generally, tags will not be written at the portal because you cannot assure speed, height, or angle of the objects and tags as they travel through the IZ. It is often impossible to read 100 percent of the cases on the pallet as it traverses the portal; therefore, it is important that you understand that cases can be identified through data aggregation as well.

When designing the portal, you must accommodate the size of the doorway. Standard dock doors are 10 feet wide (approximately 3 meters), and ramp-loading doors can be up to 14 or 17 feet wide (approximately 4 to 5 meters). For wider portals, you may need to adjust power levels and antenna types (gain) to produce a consistent read field across the whole distance.

You should understand the movement of goods to the doorway and the loading or unloading patterns of that particular facility prior to designing and installing a portal. The approximate speed of travel through the portal is another important consideration. Travel speed will help you determine how wide the antenna beams need to be to keep the tags within field long enough to assure maximum readability. You should know the maximum (or expected) amount of tags to be read per pallet, vehicle speeds and routes, and you should configure the portals accordingly. You must also recognize the physical robustness of the equipment that is going to be employed in the read IZ to ensure that the RFID equipment will not be easily damaged and will survive the environment in which it must operate.

In a portal, a circular antenna is often used, since the orientation of the tags entering the IZ cannot be guaranteed. Ensuring the proper overlap between inward-facing antennas is also critical to achieve successful reads in the IZ. The antenna fields should overlap where the tags will most likely travel. Don't forget that the width of the antenna is being narrowed as you get closer to the antenna, as discussed in Chapter 2. Therefore, the antennas should be set up for the tagged objects to travel at least 1.5 to 2 feet from the face of the antenna. If the tags were too close to the antenna mounting structure, the width of the IZ would be so small that the tags could pass outside of the coverage area.

It is prudent to angle the antennas 25 to 45 degrees into or away from the path of travel of the objects traversing the IZ. This will dramatically help you increase the likelihood of communication with tags as their time in the IZ or the dwell time of the tag is increased (see Figure 7-3). As discussed earlier, antennas with lower gains have wider beams and antennas with higher gain have narrower beams; therefore, you may want to use a lower gain antenna to achieve a wider beam and longer dwell time if the distance the antenna covers is sufficient.

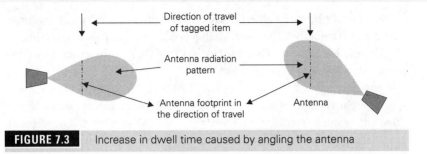

FIGURE 7.3 Increase in dwell time caused by angling the antenna

For this reason, you should use an antenna with the least amount of gain that can be used to cover the half of the door on which that antenna reads.

Post-installation Performance Measures

One of the performance measures for portals post-installation is the *percentage of tags read*, which is the likelihood of a tag on the object being read as it passes through the portal. For instance, if 100 boxes pass through the portal on a pallet, and only 80 of them are read (because products are made of RF-difficult materials), the percentage of tags read is 80 percent. This number should be set as an expectation with the client as well as a baseline for troubleshooting later.

The next performance measure is *spurious tag reads*. Spurious tag reads occur when tags are read where they should not have been. This can happen when tagged objects are accidentally read by the portal, although they are just passing by or are staged nearby. This is usually caused by unwanted reflections of the portal's antenna or excessive power settings. The system should monitor all data, and if unexpected tags are read—for instance, tags on goods staged for shipping that are not being loaded—it should filter this data out. A well-designed system will recognize the event, record the spurious read, and alert an administrator for analysis.

The following can help eliminate spurious reads:

- A new standard operating procedure (SOP) can be instituted that will make employees maintain "safe distance" from an active portal when dealing with tagged goods.
- Power in the interrogation zone could be lowered to keep the antennas from reaching distant tags.
- Directional antennas can be used in these types of installations to reduce spurious tag reads. Antennas with high gain and a well-defined primary beam will limit the coverage of an unwanted area.

- If the direction of the tagged items is known, a separate set of antennas for inbound and outbound reading can be installed, facing in or out of the warehouse, respectively. To reduce the chance of overlap dramatically, alternating dock doors could be used for shipping and receiving goods.

RFID Tunnel Installation

When building a conveyor IZ or a tunnel, remember that every tag that passes through the tunnel must be read. To achieve this result, you must understand the speed and separation of moving tags, the typical physical size of cases/units, as well as the reading environment, such as metal rollers on the conveyor. The advantage of installing a tunnel on a conveyor is a predictability of factors, such as the speed of the object passing through the IZ or the number of tags in the IZ at any given time. Another advantage is the fact that the read distance is much smaller compared to a doorway portal.

Best practices on a conveyor (in a tunnel or conveyor portal) are using the directional antennas to limit the read field to the immediate item moving on the conveyor surface and reducing power to avoid interference among other readers or reading items on adjacent conveyors.

RFID tunnels usually come preassembled for easy installation. You can simply connect them to a power and network and configure the readers with recommended parameters. If your tunnel comes without the RFID equipment, you will need to install the antennas and a reader. Tunnels usually include mounts or holders for attaching the antennas and a reader as well as space to run the cables. You should follow the tunnel manufacturer's instructions to mount and install the RFID equipment and the tunnel itself.

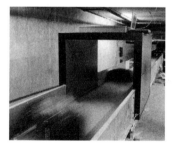

Encoding Station Installation

Encoding stations are often installed simply to meet the mandates of downstream trading partners, known as "slap and ship." A typical encoding station, as shown in Figure 7-4, includes a network-connected computer, a

FIGURE 7.4   Verification tunnel on a conveyor

barcode reader, and an RFID interrogator as a validation point for the newly applied tags.

Installing a computer connected to the back-end system allows transfer of data for the objects to be tagged. Normally, the system will have a barcode reader attached to read existing barcode information on a package. This information can be processed by the computer and used to generate the RFID-enabled smart label that will be applied on the same box. Post application, the box will be sent through a read point to ensure that the tags are still readable.

Installation on a Vehicle

A forklift truck installation is by far one of the most complex installations of all. Selection of equipment for forklift installation is critical to the success of the operation, as all equipment must meet the environmental specifications found on a forklift truck, including shock, G forces and vibration, environmental extremes such as temperature and humidity, and possible exposure to elements such as rain, wind, or snow. You must also consider how the forklifts are used in the business as well as what items they are picking up and moving.

For a proper installation of the equipment on the forklift, you must do the following:

- Use shock absorbing mounts when mounting readers.
- Mount readers close to antennas to limit the cable length (to minimize the signal loss, as discussed earlier).
- Mount the antennas discretely so that they can obtain the appropriate read but will not be subject to the damage typically associated with being at such close proximity to the objects being read (see Figure 7-5 for an example of antennas mounted on a forklift).
- Deliver power and data connections between the truck and the reader. For readers attached to the lift of the forklift, buying hydraulic hoses that have embedded wiring can be extremely advantageous to reduce the possibility of cable damage.
- Ensure operator safety by mounting the reader and the antennas out of line of sight of the operator who is performing normal functions with the forklift.

FIGURE 7.5   Vehicle-mounted antennas between forklift forks

Finding the way to tie the reader and the antennas to the back-end data system can be challenging on a forklift installation. You can often use a vehicle mount computer or vehicle mount terminal that communicates with the back-end system wirelessly. The reader will read the tags, and then pass the data to the vehicle mount computer via a serial connection, which will then filter and aggregate that data and turn it into tangible business events that can be processed by the existing warehouse management system or business transaction processing system.

A vehicle-mounted computer or terminal with a screen will also provide the driver with the information about the product he/she is supposed to pick up, product location, whether the tag on the product was read, and where the product belongs. This system ensures efficiency of workers and minimizes problems caused by faulty tags and misplaced products. An example of a computer screen in such a system is shown in Figure 7-6.

Access Control Installation

Access control points often use RFID-enabled security systems. Instead of tracking boxes through a portal or tunnel for inventory management purposes, these points are used to guarantee or refuse access to people moving in and out of sensitive business locations. By using the data collected by these systems, you can also track movement and location of personnel.

FIGURE 7.6   Screen from a vehicle computer showing graphical representation of the warehouse, including information about the product to be loaded.

Many of these systems use HF technology, but they are not limited to it. Access control points can employ the use of power locks or access gates that function similarly to a divert gate sorting boxes on a conveyor. The device responds to the data the reader has seen and the criteria defined for the object attached to the tag.

 Objective 7.3 # Grounding Considerations

Prior to plugging in readers or electronic components, you must check the grounding at the site. Three types of grounds are used: power line ground, local ground, and RF ground. When these grounds are different, potential ground loops can develop. For this reason, all components in an RFID system should be grounded to a common ground.

Prior to installation, you should stop at an electronics store and spend $5 on an inexpensive ground fault device that will indicate the condition of the power line. Simply plugging in the meter will give you a status of the line with three indication lights. The lights will turn on to indicate whether you have reverse polarity, neutral reverse, or disconnected ground. Many line conditions can be quickly assessed using this type of meter. (Never try to fix these issues yourself. Instead, call a licensed and qualified electrician to take care of the job.)

Travel Advisory
Installing surge protectors for the equipment is a good idea. Spend additional money on a UPS so that the A/C power can be filtered prior to delivery to the electronic devices.

Remember that the equipment mounted on material handling devices such as a forklift conveyor or other device should be grounded to the frame of the device.

Exam Tip

Most often you will deal with grounding the interrogator to a forklift frame.

 **Post-installation Testing**

Upon completion of an installation, you should visually inspect all power and Ethernet cables and all RF fields, and test the system to ensure that the tags are read, data transmission is occurring where it should be, and that all cables are connected properly and performing their functions. Visually inspect the RF field by introducing tagged items with properly encoded tags into the IZ, and then watch for the appropriate lights on the reader and data populating the back-end system. Move the tagged item around the intended coverage area to ensure that reads occur wherever they should.

You may need to install a status indicator with your portal to indicate when it is "live" or reading tags so that facility personnel can avoid walking through that IZ with tagged items that are not supposed to be read. Such rules may indicate the need for a change in operational parameters for that facility.

Once the system is fully installed, you should officially commission the system by giving the client a demonstration of its abilities with tagged goods as they are processed through the facility. Once the customer is satisfied with the completion of the installation, obtain a signature of an authorized person indicating acceptance of the installation.

 **CHECKPOINT**

✔**Objective 7.1: Preinstallation**   Prior to installation, you should determine whether you have proper equipment and safety devices ready for the installation. Make sure that the RFID equipment functions properly in accordance with the manufacturers' specifications and will function as desired if installed in locations the way it is designed by a system architect. Check whether the appropriate cables and wiring are installed in the right places (and that they comply with industry standards). Ensure that adequate space is afforded to run the equipment and that the power line is properly grounded.

✔**Objective 7.2: Hardware Installation**   All static hardware must be properly secured by adequate mounts, and grounded and protected from external damage caused by personnel, vehicles, or environmental conditions (using special stands, bollards, and NEMA enclosures). Hardware must have enough space for access to connections and cabling as well as for ventilation. If you cannot ensure that the equipment will be used in a dry, stable environment, you should use protective NEMA enclosures. Mobile equipment must be connected properly to achieve the desired mobility. When you are installing any electronic equipment, first connect all additional devices to it before powering it.

✔**Objective 7.3: Grounding Considerations**   Three types of grounds are used: power line ground, local ground, and RF ground. When these grounds are different, potential ground loops can develop. All components in an RFID system should be grounded to a common ground. Use surge protectors or uninterruptible power supplies (UPS) for the equipment. The equipment mounted on material handling devices such as conveyor or other stationary devices as well as a vehicle such as a forklift should be grounded to the frame of the device. In areas prone to lightning, install lightning protectors.

✔**Objective 7.4: Post-installation Testing**   Post-installation, visually inspect all power and Ethernet cables, visually inspect all RF fields, and test the system to ensure that the tags are read, data transmission is occurring where it should be, and all cables are connected properly and performing their functions.

REVIEW QUESTIONS

1. Which of the following is a concern when installing a passive ultra high frequency (UHF) RFID system that complies with all regulations?
   A. Electrostatic discharge
   B. Loss of hearing due to UHF exposure
   C. Interference with local FM radio broadcast
   D. Interference with adjacent interrogation zones

2. When an interrogator is installed on a forklift, which of the following actions should be taken?
   A. Since this is not an AC system, it is not necessary to ground the connection.

B. Connect the interrogator ground to the negative terminal of the battery.

C. Connect the interrogator ground to the antenna ground plane to establish common ground reference.

D. Connect the interrogator ground to the forklift frame.

3. You are designing an RFID installation for dock doors in a facility. The facility is located in an area with high wind and heavy rains. Which of the following is the best solution to protect interrogators?

A. Install dock shelters on dock doors.

B. Place the interrogators in NEMA-4 enclosures.

C. Place the interrogators at least 5 feet from the dock doors.

D. Put heavy-duty plastic bags over the interrogators.

4. An interrogation zone on a conveyor is capable of reading only 10 tags per second. However, management decided to increase the speed of the conveyor, so that the interrogator would have to read 12 tags per second. What is the best solution for this problem?

A. The speed of the conveyor cannot be increased.

B. Install another antenna on top of the conveyor.

C. Increase the distance between boxes.

D. Tilt the antennas on the opposite sides of the conveyor 45 degrees, one toward the incoming tagged boxes and the other toward the outgoing boxes.

5. A warehouse has 60 adjoining dock doors that are all equipped with UHF interrogators, which interfere with each other. Which of the following is the best way to resolve this interference?

A. Configure even interrogators to operate at high frequency and odd interrogators at ultra high frequency.

B. When one interrogator is turned on, others have to be turned off.

C. Install RF shielding around each interrogation zone.

D. Tilt the antennas toward the dock doors.

6. At the dock door portal installation, which of the following is the best way to protect antennas from damage by forklift?

A. Enclose antennas in thick metal boxes.

B. Install padded bollards by the antennas.

C. Enclose antenna in NEMA-4 enclosure.

D. Use only the original manufacturer's stand.

7. Why should you prevent electrostatic discharge (ESD) when handling electronic equipment?

   A. ESD shocks people.

   B. ESD may damage delicate electronic equipment.

   C. ESD is not a problem with RFID systems.

   D. ESD creates a ground loop.

8. What are the best practices for interrogator and antenna installations? (Select two answers.)

   A. Make sure that the interrogator has LED lights.

   B. Power up the interrogator and ensure that it is working before connecting antennas to it.

   C. Test the interrogator and antenna before installation.

   D. Test the interrogator and antenna after installation.

9. To cut down on installation time, which of the following actions should be taken? (Select two answers.)

   A. Work fast and do not let anything disrupt you.

   B. Preassemble and test all equipment prior to installation.

   C. Ask facility employees to help you with reader configurations.

   D. Bring spare parts onsite, just in case.

10. What is the best practice when installing interrogators?

    A. Leave 5 to 6 inches of space around the interrogator for cables, access, and ventilation.

    B. Encase the interrogator tightly in a metal box, so that it cannot move even if not attached by screws.

    C. Never leave the interrogator without a proper enclosure such as NEMA.

    D. Use only shielded cable conduits through which to run the antenna cables.

REVIEW ANSWERS

1. **D**   When installing a passive UHF RFID system that complies with all necessary regulations, the main concern is the interference with adjacent interrogation zones.

2. **D**   When an interrogator is installed on a forklift, connect the interrogator ground to the frame of the forklift.

3. **B**   You are designing an RFID installation for dock doors in a facility. When the facility is located in an area that gets high wind and heavy rains that sometimes gets inside the facility, it is best to place the interrogators in NEMA-4 enclosures.

4. **D**   If you need to increase the speed of the conveyor so that the interrogator would have to read 12 tags per second as opposed to original 10, you should tilt the antennas on the opposite sides of the conveyor 45 degrees, one toward the incoming tagged boxes and the other toward the outgoing boxes, to widen the read field and therefore increase a tag dwell time.

5. **C**   If a warehouse has 60 adjoining dock doors that are all equipped with UHF interrogators that interfere with each other, you should install RF shielding around each interrogation zone.

6. **B**   At the dock door portal installation, the best way to protect antennas from damage by forklift is to install padded bollards by the antennas.

7. **B**   Electrostatic discharge (ESD) has to be avoided because it can damage delicate electronic equipment.

8. **C** **D**   The best practices for interrogator and antenna installations are to test the interrogator and antenna before and after the installation.

9. **B** **D**   To cut down on an installation time, you should preassemble and test all equipment prior to installation and bring spare parts onsite, just in case any equipment fails or gets damaged.

10. **A**   The best practice when installing interrogators is to leave 5 to 6 inches of space around the interrogator for cables, access, and ventilation.

Testing and Troubleshooting

ETA	NEWBIE	SOME EXPERIENCE	EXPERT
	4 hours	2 hours	45 minutes

The methods used for testing and troubleshooting a failed RFID installation are similar to those used when troubleshooting classic IT systems. You always start with the easiest data points to verify at the physical device, and then you work your way out to the more complex connections to back-end systems.

Objective 8.1 General Troubleshooting Process

When starting the troubleshooting process, you must clearly define the failure before you do anything else. This will provide the starting and ending points for this process. Once you understand the problem as interpreted by the user, you can work to resolve that specific issue.

Remember that getting the system running and finding the source of the failure are usually two very different things. Ensure that you have identified the actual source of the trouble so that you can devise a way to prevent this problem from occurring again in the future. This is called *root cause analysis*, and it's often forgotten once the system is running again. When this crucial step is missed, further problems can arise from the problem you thought you solved.

Follow these steps when troubleshooting:

1. *Reboot the system.* Before you begin any troubleshooting, reboot the equipment in the interrogation zone (IZ). This simple step will often resolve issues brought on by power surges or incomplete commands issued by the back-end system.

Exam Tip

Remember that rebooting is often the easiest way to fix the problem and should be one of the first steps in the troubleshooting process.

---

2. *Start with easy stuff.* Always begin your investigation by looking at issues that are easily discovered and verified. Check power connections at the reader as well as at the source, including any power converters or conditioners. Check cables for damage in the cables themselves or at

the connector, because excessive bending or other signs of possible cable wear might lead to signal degradation and/or signal interference. Check all physical connectors to see that they are all properly seated in their sockets (don't forget antenna connectors and I/O connectors), and be sure to look at the power indication lights on all the equipment. This step may seem obvious, but many hours of troubleshooting have been wasted only to discover a blown power supply or a disconnected power cord.

3. *Check whether and where the tags are seen.* Once you have verified that power is running to all devices and components, test to see what happens when tags are present in the read area. Most interrogators have several indicator lights that show not only the reader status but also whether the tags have been recognized and interpreted. If complex triggering devices are installed within the IZ, you can disconnect the interrogator from all other devices, connect to it with your laptop, and then perform this test using the manufacturer's software provided with the interrogator. If you see the tags, you know the interrogator is not the source of the issue; if you don't see the tags, look closely at the antenna connections and cables, interrogator configuration, and finally the interrogator itself for possible replacement.

Travel Advisory

When replacing hardware, make sure that you turn off the whole system and disconnect the power running to a device. Never disconnect an antenna from an interrogator when it is in use—even if the antenna is not in use but the interrogator is turned on. When you are installing a new device as a replacement, follow the same rules. Make sure that you connect the devices to the power as the last step after you secure all other connections (antennas, network, I/Os, and so on).

4. *Test all triggering and feedback devices.* If the tag was successfully read when you manually triggered the interrogation through the manufacturer's software but it is not triggered (or at least you cannot see any indication if it is) when the tagged product goes through the IZ, you need to check your triggering and feedback devices. Are the light break sensors properly aligned, connected, and configured? Is the

reader set up for polling for tags after being triggered by an external device? Are the presence detectors lighting the appropriate lights when materials are in the interrogation area? Does the light stack show any lights at all? If the equipment acts like it is functional, but no indicator lights are lit, check for appropriate power and connections and verify whether the light bulbs are functional or need to be replaced.

5. *Check network communications.* Once all the local devices belonging to the IZ have been verified, look at network connections to and from the zone. Connect a PC to the network on which the interrogator is connected. Ping the interrogator (see the following "Travel Assistance"). If you get a response, you can rule out connectivity and network configuration issues. If you do not get a response, you can check the network configuration for appropriate settings of the IP address, subnet mask, and gateway. You should also perform this test for any network connected devices such as Programmable Logic Controllers (PLCs) instead of path loss contour devices.

Travel Assistance

How do you ping a reader? If you have a computer networking background, you know how to do this, but here's a brief description for the rest of us. Open a command prompt in your computer: choose Start | Run, and then type **cmd** and press ENTER. A window opens, where you will see a path to your current location. Type in the following: **ping 192.168.xyz.xyz** (the *xyz* number is the IP address of your reader—don't enter xyz.xyz!) and press ENTER. If you do not know the reader's IP address, you can find it through the hyperterminal via a serial connection or through your network router interface. If you get a reply to the ping, this indicates that your reader is communicating over the network. If you see "Request timed out," you will have to troubleshoot the network settings and connections.

6. *Check middleware.* If all the local devices can communicate across the network properly but your reads are not registering in your back-end systems, you need to look at the other end of these connections. If a middleware client is placed behind these readers, make sure that it is running. Just as you did with the IZ equipment, reboot it! A full server reboot may not be feasible, but you can certainly restart the service

controlling this package to ensure that no corrupt data has interrupted it. If this resolves the problem, be sure to check the log files to see what the last event prior to the failure was so you can perform root cause analysis.

7. *Check enterprise resource planning (ERP), warehouse management systems (WMS), and Transaction Processing (TP) systems connections.* After you have tested the middleware and verified that it functions as it should, check the connections to the back-end system to ensure that they are functional. Typically, installed systems will have a status monitor that indicates failure should one exist. Sometimes you can run a test script that will emulate data coming from the reader to ensure that the back-end system shows the proper results. If you encounter problems, restart the service (reboot) and see if the problem persists. As before, always check the logs to find the last event prior to failure to perform root cause analysis.

After completing these steps, most problems should be resolved.

Exam Tip

Be aware that if the RFID equipment follows the RF emission regulations, the damage it can cause to a human body during operation and handling is not likely to be due to radiation but rather due to electric shock.

 **Objective 8.2**

Interrogation Zone Troubleshooting

Following are common problems that you might encounter during system deployment. Most problems will be generated by improper hardware and software setup in the IZs or by improperly tagged products. Damage to hardware as well as tags themselves are also often causes for failures. Let's look at some specific problem examples and troubleshooting techniques that will help you resolve these potential situations.

An infinite number of problems can be generated by a failure or improper setup of the IZ. Following are some of the most common problems that you need to check and information about how to troubleshoot the problem.

No Reads

The IZ doesn't show any reads although tagged product is going through the IZ. What questions should you ask?

- *Are all the components powered on?* You should see various status lights blinking or turned on. Check the interrogator lights, lights on the power supply, and lights on the peripherals. If any of these lights are not illuminated when they should be, check to ensure that the connections are secure and check the condition of the cables. Make sure everything is tight and properly secured, that cables are not bent more than they should be, and that the connectors are not damaged.

- *Is the reader configured properly?* Prior to the reader being able to read tags, several settings must be correctly configured according to the manufacturer and type of the reader. Check the following:

  - *Is the world region set?* Some readers will not read tags until you set them up for the appropriate world region and country of operation. These settings concern various performance restrictions around the world.

  - *Is polling enabled?* Make sure that polling is enabled and set up for certain time intervals, continuous polling, or waiting for a trigger from an I/O device according to your needs. The default setting may prevent the reader from polling (reading) for tags.

  - *Are the correct tag protocols defined?* Is the reader set up for Electronic Product Code (EPC) Class 0 and Class 1, and do you need to read Gen 2 tags? Or is the opposite required? Is your reader set up for reading EPC tags, while you are running International Organization for Standardization (ISO) tags through your IZ? Make sure that you set up your reader for the appropriate protocol. Try to avoid choosing support for all protocols because that would slow down your read rates. Use this only if you are truly reading all of the reader's supported protocols in your operation.

- *Are the antennas defined correctly in the reader?* Most readers usually automatically recognize which antennas are connected and automatically enable them. However, if your reader drives the antennas for two or more IZs, verify that the appropriate antennas for that particular zone are enabled and set with correct output power settings.

- *Is the read/write function of the reader set?* Some readers can be set for read only, write only, or both, so make sure that the appropriate function is not disabled.

- *Is the network configuration set up correctly?* Your reader may read tags but, due to improper network configuration, you are not seeing the reads in your middleware or back-end system. Make sure that the Dynamic Host Configuration Protocol (DHCP) is turned off or on, depending on the network. If you are not using DHCP, you must use the appropriate IP address to communicate with the reader. If you are using DHCP, the IP address will be dynamically assigned to the device by the network.

- *Is the antenna orientation correct?* If using linear antennas and single dipole tags, ensure that the tags are parallel to the antennas.

  - *Can the antenna beam reach the tag?* Make sure that you have correct power settings for the antenna to ensure appropriate RF coverage of the IZ.

  - *Can the tag be seen from all positions on the product?* With dense or RF-difficult products, you may need to angle the antenna or add additional antennas to ensure sufficient coverage of the IZ and the product to achieve reads. Your tags might have been on the bottom of an RF-opaque product, for example, so antennas reading from the top couldn't reach the product.

Less Reads than Tags

The amount of tags passing through the IZ doesn't match amount of tags read (but tags are functioning). If the amount of tags read is smaller than the amount of tags traversing the IZ, consider the following questions:

- *Is the expectation of reading all tags in the zone realistic?* The answer may depend on the type of product, amount of tags, spccd the tags are moving through the zone, as well as the IZ configuration such as number of antennas and reading settings.

A full pallet of goods may not always read at 100 percent. This happens often with dense or RF-difficult products such as cans or metallic components. Additional antennas aiming from the top of the pallet may provide additional reads, as may pallet rotation. Alternatively, you can compare the case tags with the pallet tag, and if the product stretch wrap is untouched or the case count is confirmed, you can handle the pallet as 100-percent read.

- *Are too many tag protocols enabled?* Too many protocols enabled will slow down the reader because it has to check the tag for every protocol until it finds the right one with which to communicate successfully. This can cause missed tags, so you should enable only the protocols required for the operation.

- *Are the goods in view of the antenna for a long enough period of time?* If many tags are moving quickly through the IZ, the reader may not be able to read them all. This can be caused by a number of protocol issues already discussed, or it could be caused by other processes, such as anti-collision or dense reader mode. The solution is to increase the tag's dwell time by angling the antennas into the tag's path of travel. It may be necessary to place more antennas along the path to increase IZ coverage.

- *Are potential sources of interference located around the IZ?* Interference can cause missed reads. Check for wireless communication systems and wireless networks as well as metal objects in the area that can cause signal reflections and multipath interference.

More Reads than Tags

If the number of tags read is larger than the amount of actual tags passing through the IZ, ask the following questions:

- *Does the data format of the additional tag reads match data encoded to other tags?* If the answer is yes, you are probably experiencing stray reads. These reads occur when a product is passing by the IZ and due to improper power setting or reflections in the IZ, the antenna registers the tag although it shouldn't. Similar situations can occur if the power settings at the adjacent IZs such as dock door portals or conveyor portals are too high and the antenna beams are overlapping to adjacent IZs. To solve this problem, you should lower the power settings, tilt the antennas away from the passing object or use antennas with smaller gain, or use shielding between each IZ.

- *Is the data format inconsistent with any other data formats used on your tags?* Does the data look totally "nonsensical"? You may have experienced a ghost read. Ghost reads are caused by RF noise in the environment that the interrogator can sometimes interpret as a tag data. Although the Gen 2 protocol has a technique for decreasing the ghost reads by format verification, ghost reads may still appear from time to time.

Objective 8.3 Tag Failure Troubleshooting

Tag failure can be one of the most difficult problems to troubleshoot in an installed system. Tags fail for many reasons, such as improper application speed, improper application pressure, electrostatic discharge (ESD), tag abuse, incorrect tag programming, and poor tag quality. Following is a discussion of a few of these scenarios and suggestions of things to check to isolate the source of the problem.

Tag Not Recognized

Tags are being read but not recognized by the inventory system. Ask the following:

- *Is the format of the tag data expected by the system?* A 64-bit value assigned to a 96-bit memory space for an EPC value will give the appearance of a corrupt tag. Reader output parameters should be checked to ensure the format of the messages they send is what is expected by the data processing system.

Travel Assistance

EPC Generation 1 tags used to support a 64-bit EPC number, but Gen 2 uses 96-bit EPC and supports up to a 256-bit EPC number that may be used in the future.

- *Does some data on the tag appear to be invalid?* This tag may have slipped through the encoding procedure. Manufacturers place test data on the tags to validate their operational capacity prior to shipment, and this may be the data you are seeing if a tag slipped through your process unencoded. There is also the possibility that a certain part of the EPC number is incorrectly encoded or corrupted. For instance, if there is a corrupted EPC header, the data read from the tag will not make any sense.

Conveyor Reading

On a conveyor, the amount of tags moving through the IZ doesn't match the amount of tags read. Ask the following:

- *Are the tags placed properly on the objects to which they are attached?* Tags may have fallen off the product. You must also ensure that the tags are in the proper place. The product was tested for the best tag placement, so compare the best tag location with the actual tag location. If the tag is placed in a location that is not ideal, it may get detuned and missed during interrogation.

- *Are these quiet or damaged tags?* Depending on the testing and verification procedures, the product could have been tagged with a quiet or damaged tag. A tag could have also been damaged during its application, or during its life by the conveyor, other products, forklifts, or other means.

  Tags are often damaged by ESD, which can result from accumulation of static charges. Static charges build up in environments with very low humidity; around various types of plastics, carpets, and other static susceptible materials; and by traction. Static charges are created by peeling off the label from the backing, walking on a carpet, traction of conveyor belt bearings, stretch wrapping, and many other activities. To protect the tags from ESD, you should store them in ESD-protective bags (conductive bags) and take other ESD protection and avoidance measures as discussed in previous chapters.

Tags can be also be damaged by extreme temperatures. Tags on a roll have a lower storage temperature than tag operational temperatures. When tags are on a roll, increased temperature will increase the pressure on the tags as the roll tries to expand. This can cause broken connections between the ICs and the tag antennas. You should follow manufacturer's guidelines regarding storage temperature and tag operating temperature, which is on average from about −20 to 70 degrees C (−4 to 158 degrees F).

You must determine whether the damage to tags occurs at a common place and how it occurs and possibly try to eliminate this point or add a step for exception processing.

- *Is the tag oriented properly to the antenna?* As mentioned earlier, if the IZ uses a linear antenna and the tags are single dipole, you must ensure that they stay in proper orientation while traversing the IZ. At some point in the process, the tagged cases may get flipped or turned around, and that could cause the changed orientation. You may need to eliminate this issue or implement circular antennas to ensure the reads regardless of the tag orientation. Sometimes the orientation issue doesn't lie within the horizontal or vertical orientation but with the angle toward the antenna. If the tags are placed at a 90-degree angle to the antenna, they are quite unlikely to be read (depending on the type of tag and the RF propagation in the area). Combining more antennas and possibly tilting them in various directions can help solve this problem.

- *Has the tag been properly selected for this product?* This problem can occur mainly if you are using the same tags for a wide range of products. If the product is only cereal, for example, the best performance is probable along with 100 percent reads. But if you are using the same tag for tagging and shipping cans of food, many reflections and detuning will occur, which will result in a much lower read percentage. You often cannot change the tag selection, but you can at least revise the tag placement or increase the performance by adding an insulating or air gap creating layer to reduce or eliminate detuning.

Hardware and Software Troubleshooting

Objective 8.4

W e've discussed what to check and how to determine causes of problems with a device as well as what to do if performance is worse than expected. These problems are usually caused by improper configuration. Now let's look at hardware and software issues that may cause less than ideal performance of the device.

Firmware Upgrades

Known software issues within the hardware can often be addressed through firmware upgrades from the manufacturer. However, these updates should be applied only if they address a specific need. Every firmware update for RFID interrogators, RFID printers, or any other device will include a text file that describes the issues it specifically addresses. Manufacturers use these upgrades to fix problems identified in units after they have been released.

Firmware updates can also be used to increase the functionality of components. For instance, many readers and printers could be firmware upgraded to support Gen 2 tags. In case the firmware introduces a new feature or support for a new technology standard that you will be using, an upgrade will save you money and time you would spend obtaining new devices.

It is always a good idea and best practice to make sure that all similar devices are running on the same version of firmware. This can help ensure consistency, both in performance and in maintenance. Different versions of firmware may cause discrepancies in how various readers perform, even if they are the same model. As new firmware is released, it may cause the reader to run faster due to enhancements, for example. To obtain uniform performance, test a firmware on a sample reader to see that it works as expected, and then deploy it to all readers of that type.

When building a maintenance plan for your equipment, you should always have a documented procedure to load the specific firmware your company uses on the device before deploying or replacing it. This way, all of the options in your configuration are sure to be available, and all the features you need are where your technicians expect them based on the documentation.

Be careful not to load a version of firmware simply because it is the latest available. The rule here is "very old and very simple." Or, "If it isn't broken, don't fix it!" If you decide to try out new firmware, always test it on an isolated device designated for testing to ensure that it is not going to cause problems. Subtle changes in an upgrade can wreak havoc on an implementation if the upgrade is not implemented properly. For instance, command sets could change between versions, rendering installed communications useless and re-quiring extensive re-engineering of software components. For these reasons, always test new versions thoroughly in every use scenario in the company (dock door configurations, conveyor configurations, and so on) before con-sidering full-scale rollout.

Network

Network communication problems can dramatically impact the performance of an installed RFID system. If you recorded a baseline of performance during installation, you'll find it beneficial to check to baseline to determine the average response time then. Comparing the original response time to the current situa-tion may indicate the network is experiencing higher network traffic, causing a slow response from the RFID installation.

Software

When all of the physical devices and the network have been verified and are working, proceed to checking the software. Most middleware platforms include some sort of status monitor that can be used to verify at a glance that the pro-gram is functional. Sometimes a test script is included to allow you to input a known data stream and expect a fixed value to be output. If the script produces the expected value, the problem may be somewhere else. You should investigate the connection from the middleware to the legacy application.

All software can be restarted (like step one with hardware) by rebooting it. When a service appears to be locked up or not responding to regular input, re-starting it is the first step. If this does not resolve the issue, check the error logs for a logged failure. Even a recent administrator password change can cause is-sues for services and will appear as such in the error logs.

Certain errors can be also cleared by restoring the reader's default settings. Some readers have a safe mode that does not allow reading tags or other regular operations but restores default settings and tests various reader functions.

CHECKPOINT

✔**Objective 8.1: General Troubleshooting Process**   When troubleshooting the IZ, follow these steps:

1. Reboot the equipment in the IZ.

2. Check power connections including any power converters or conditioners, antennas, and network and I/O connections.

3. Check if and where the tags are seen. If tags are not seen, check antenna connections and cables, interrogator configuration, and the interrogator.

4. Test all triggering and feedback devices.

5. Check network communications and proper network configuration.

6. Check middleware if all the local devices can communicate across the network properly but your reads are not registering in your back-end systems.

7. Check ERP, WMS, and TP systems connections.

✔**Objective 8.2: Interrogation Zone Troubleshooting**   When an IZ doesn't show any reads when a tagged product is passing through, the failure may be caused by a disconnected power source, improper reader configuration (a world region, enabled polling, enabled read/write, correct air interface and data protocols, network information), antennas incorrectly defined in the reader, improper antenna configuration, or improper tag orientation related to an antenna.

If the amount of tags read is smaller than the amount of tags traversing the IZ, this could be caused by too many protocols enabled (which leads to lower read rates), short tag dwell time, interference, improper tagging, or incorrect tag types for the given product.

If the amount of tags read is larger than the amount of actual tags passing through the read zone, you are experiencing stray reads from tags near the IZ. If the data format of the additional tag read does not match data encoded to other tags, the phenomenon is called ghost reads.

✔**Objective 8.3: Tag Failure Troubleshooting**   When the amount of tags passing through the IZ is higher than amount of tags read, it could be caused by improperly tagged items. The product should be tested for the best tag placement, so you need to compare the best tag location with the actual tag location. If the tag is placed into a less-than-ideal location (for instance, touching a RF-difficult product such as metallic or aqueous items), it may get detuned and missed during interrogation or torn off during its move through the supply chain. Other problems can be caused by using improper types of tags for a product. Proper tags should withstand the environmental conditions and handling to which the product is exposed. In addition, you should use a tag that is tuned to operate optimally when placed on certain materials (such as tags specifically made for metal, glass, or other materials).

✔**Objective 8.4: Hardware and Software Troubleshooting**   To determine why the hardware is not functioning properly or doesn't work at all, you must check following things:

- Power connections at the reader as well as at the source, including any power converters or conditioners.
- All cables for proper connectivity or damage on the cable or at the connector, excessive bending, or other signs of possible corruption.
- Proper function of triggering and feedback devices. The reader may be set up for polling for tags after being triggered by an external device, but if the device doesn't function, the reader doesn't get the trigger.
- The lights on the devices and light stacks. If the equipment acts like it is functional, but no lights are lit, check for appropriate power and connections and also verify whether the light bulbs are functional or they need to be replaced.
- Network connectivity. Connect a PC to the network on which the device is connected. Ping the device. If you get a response, you can rule out connectivity and network configuration issues. If you do not get any response, you should check the network configuration.
- Firmware updates can be used to fix software issues in the hardware, increase the functionality of components, and provide support for new features (such as support for Gen 2 tags).

REVIEW QUESTIONS

1. In a shampoo-packaging facility, shampoo bottles are packaged into cardboard boxes. An automated printer applicator is used to print and apply smart labels to the cases; however, some of the boxes are arriving to the palletizer without labels. What is the best way to correct this problem?

   A. Change label dimensions.

   B. Slow down the conveyor belt and speed up the label applicator.

   C. Install an IZ downstream from the printer applicator to verify the labels and route the untagged boxes to exception processing.

   D. Have a person verify that each box has a label.

2. A company has been using an EPC Class 1 Generation 1 RFID system, but one of the suppliers recently announced that it will use EPC Generation 2 tags on its shipped products. What is the best solution for the company if it wants to read the Gen 2 tags?

   A. Replace the current interrogators.

   B. Change the tag type to Gen 2 in the interrogator configuration.

   C. If possible, upgrade the interrogator firmware to support Gen 2.

   D. Modify the middleware to issue Gen 2 commands to the interrogators.

3. In an existing UHF RFID system, the interrogator is connected to three antennas. The system has been working properly, but recently no reads were provided by antenna 3. Which of the following actions should the RFID technician take first?

   A. Check the interrogator configuration and make sure that antenna 3 is enabled and set up for interrogation.

   B. Check whether any RF blocking obstructions are in front of antenna 2.

   C. Verify that the interrogator is powered up properly.

   D. Check the antenna cable and make sure it is connected properly.

4. When pallets with frozen food are received from a refrigerated truck, a high rate of misread tags occurs at the dock door portal. Which of the following is most likely to cause these misreads?

   A. There is a condensation on the tags.

   B. The tags on the pallet are not designed to operate when the temperature is below freezing.

   C. Some of the tags on the pallet are damaged due to the freezing temperature.

   D. All of the above.

5. UHF RFID inlays are attached to reusable plastic containers that are used in a plant for transporting various parts. A high percentage of inlays are failing during normal operation with antennas set at maximum allowed power. Which of the following reasons would most likely cause the failure?

   A. The plastic containers generate ESD, which damages the inlays.

   B. The water spray used to clean the containers is damaging the inlays.

   C. Hot air at 130 degrees F (55 C) blown on the containers is damaging the inlays.

   D. The inlays may be damaged by high power from the antenna.

6. In order to be protected from electrostatic discharge, how should the smart labels be stored?

   A. In an area with high humidity

   B. In conductive bags

   C. In refrigerated cabinets

   D. In an area with good air circulation

7. At a dock door portal, often more read tags are registered than the actual number of tags that went through the IZ. However, the additional numbers do not match the format or value of any currently encoded numbers. What is happening in the IZ?

   A. Issue with reader firmware

   B. Ghost reads

   C. Incorrectly encoded tags

   D. Stray reads

8. During troubleshooting of the IZ, the technician sees that the power and network lights on the interrogator are on but the reader does not read any tags or respond to triggering devices. Which of the following is the first step you should take when troubleshooting this situation?

   A. Replace the antenna cable.

   B. Replace the interrogator.

   C. Ping the interrogator.

   D. Reboot the interrogator.

9. In a distribution center, 20 dock doors are placed in a row. The interrogators at some dock doors are reading tags passing through adjacent dock doors. Which of the following actions should be taken to solve this problem? (Choose two answers.)

   A. Use antennas with higher gain to focus the beam and eliminate interference.

   B. Adjust power settings at the interrogators that read tags at adjacent dock doors.

   C. Replace the interrogators that are reading the wrong tags.

   D. Install shielding material between adjacent dock doors.

10. A newly installed interrogator is generating constant errors while attempting to encode tags. Which of the following is most likely the cause of the problem?

   A. All the RFID tags are quiet.

   B. The interrogator antennas are not pointing in the proper direction.

   C. Network problems are preventing proper communication with the interrogator.

   D. The interrogator configuration parameters are not set to match the tag parameters.

REVIEW ANSWERS

1. **C** In a shampoo packaging facility, where shampoo bottles are packaged into cardboard boxes that are tagged by an automated printer applicator, some of the boxes are arriving to the palletizer without labels. The best way to correct this problem is to install an IZ downstream from the printer applicator to verify the labels and route the untagged boxes to exception processing.

2. **C** If the company wants to read the Gen 2 tags, it should first try to upgrade the interrogator firmware to support Gen 2. This option is supported by most of the interrogators on the market; however, if the upgrade is not possible, the purchase of new interrogators will be necessary.

3. **D** To start troubleshooting this problem, the RFID technician should first check the antenna cable and make sure it is connected properly. It may have been disconnected accidentally or on purpose by personnel. The interrogator has to be powered, since the other two antennas are

reading, and antenna 3 has to be set up properly because it was working earlier; therefore, A, B, and C are incorrect.

4. **A** The most likely cause of these misreads is the condensation on the tags, which can reduce the tag read range. Tags applied to frozen food or food intended for freezing are specifically designed for this purpose and are able to survive low ambient temperatures.

5. **A** The failures are most likely caused by the ESD generated by the plastic containers, which damages the inlays. The ESD on the plastic containers can be generated by traction of the containers on the conveyor or when being handled during their transport.

6. **B** To be protected from ESD, smart labels should be stored in conductive bags.

7. **B** When the tag ID numbers do not match the format or value of any currently encoded numbers, the problem is most likely a ghost read.

8. **D** The first step to take when troubleshooting this situation should be to reboot the interrogator, which often fixes the problem.

9. **B** **D** When the interrogators at some dock doors are reading tags passing through adjacent dock doors, you should adjust power settings at the interrogators that read tags at adjacent dock doors and install shielding material between adjacent dock doors.

10. **D** When a newly installed interrogator is generating constant errors while attempting to encode tags, the interrogator configuration parameters are not set to match the tag parameters. For instance, the air interface or data protocol does not match.

Site Analysis

ETA	NEWBIE	SOME EXPERIENCE	EXPERT
	4 hours	2 hours	45 minutes

This chapter discusses the site survey, the planning that goes into the site survey, and various elements that you need to examine when performing the site analysis, including anomalies for the site. It also provides guidance regarding generating the site survey report.

When performing a site analysis, keep in mind where and how you will install the interrogation zones (IZs). Because each facility has unique characteristics, your considerations for every IZ will vary. You'll encounter differences in each site's workflow; differences in how personnel, material, assets, and other objects move about the site; and differences in the items you track between one site and the next, even if the sites are within the same implementation project or within the same enterprise-wide implementation across multiple facilities.

During the site analysis, you will gather all the information you need for design and deployment of the RFID system. It is critical that you get information for all the systems with which you will be interfacing for your implementation. So, for example, if you are going to connect to a warehouse management system, collect every piece of documentation you can about it. If you can get the phone numbers of the people who installed this system, document them in case you need more information. You must also obtain the contact information for the individuals who are responsible for the back-end system with which you will be integrating the RFID system. Typically, sites that are ISO 9000 certified have standard operating procedures already documented. You need to familiarize yourself with these procedures and collect information about how problems are handled; for instance, when machinery breaks, what is the contingency plan? For those sites that are not ISO 9000 compliant, you'll need to develop your own documentation.

Once you understand the characteristics of the site and its environment, including the RF characteristics, the business processes, and the flow of the products, you can predict the performance of the RFID solutions you plan to deploy.

 **Objective 9.1** # Infrastructure Assessment

Before you start working on RF site characteristics (usually the focus of most people who try to sell spectrum analyzers or expensive consulting hours), you must consider the environmental characteristics as well as the infrastructure of the site. These characteristics can "make or break" your system implementation.

Environmental Characteristics

Understand the environmental conditions of the site, because your system must be designed not only to survive but also to perform optimally in these conditions. You should be aware of such conditions as temperature extremes, levels of humidity, and possibility of water intrusion, electrical shock, and vibration.

Temperature Considerations

Consider the following: Is the installation in a temperature-controlled facility? To what temperatures are the equipment exposed during its operation? If the system will be deployed outside—for instance, in a shipping yard—electronics may be required to withstand temperatures anywhere from 25 to 45 degrees (F) higher than the ambient temperature at the site. This might not be a problem in an air-conditioned facility, but it can be troublesome during summer if the facility is not temperature-controlled, because you may exceed the maximum operating temperature recommended by the equipment manufacturer. Be aware that if the maximum operating temperature for your equipment is up to 140 degrees, with the ambient temperature reaching 110 or 115 degrees, the insides of the cases containing the electronics can reach 140 degrees or more. Computers and computer-related equipment are very sensitive to extreme temperatures and will start degrading performance or shut down as they overheat. If your equipment is often overheated, the hardware's overall lifespan will decrease and problems with reliability can come to the fore.

Humidity, Dust, and Chemicals

You must consider the types of operations that take place in the facility. Is this a "clean" facility, such as research facility, laboratory, hospital, or electronics assembly plant? In this case, consider your presence during the deployment as well as how your equipment will affect this environment once installed and operational. If this is a "dirty" facility, you'll look at factors that may affect your equipment after it is installed, such as the levels of humidity experienced in the facility, chemicals that are used in the process, or the potential for large amounts of industrial dust. (For instance, in a paper manufacturing plant, you will encounter large amounts of paper dust.) If your equipment has open vents and fans, the dust will be sucked into the heat sinks of the electronics, so you might have to add filters to the fans to reduce some of that dust prior to its invading the equipment.

Exam Tip

Know that you may need to mount some of your equipment in uniform enclosures (for instance, NEMA enclosures) to protect it from the environment to keep out water, humidity, and dirt, but you may also use temperature-controlled or explosion-proof enclosures.

Local Lingo

Dirty facility   A *dirty facility* may be not only physically dirty—full of dust, lint, or chemicals—but can also be dirty as related to RF signals. A facility in which many RF-emitting devices operate can have high levels of RF noise and interference, which provides difficult conditions for RFID implementations.

You should examine the facility environment not only because of the possible damage that it could cause to your equipment, but also to make sure that the RFID system installers will be ready to work in such space. How you mount the equipment is just as important as the equipment you select for these sites. Of course, the facility type will determine the equipment you bring with you and the clothing and breathing protection equipment you wear. (You won't wear your Hugo Boss suit during a site survey or installation at a warehouse, because you'll undoubtedly be crawling around dusty conveyors and walk around the dirty facility. On the other hand, if you're doing a site survey or implementation in a hospital, you won't want to wear steel-toe boots, an old T-shirt, and sturdy jeans.)

Vibrations

Make sure that you identify all sources of vibration that will affect your equipment. Obviously, if your RFID equipment will be mounted on mobile assets, it will experience vibration. If you are attaching an RFID interrogator to a forklift, for instance, you know that it will have to withstand the vibrations as the forklift rolls across the floor.

Some causes of vibrations might not be immediately obvious. Check for conveyors, manufacturing lines, and heavy machinery inside the facility, as well as rail lines, highways, or airports outside the facility in the near proximity. In some manufacturing facilities, rail lines literally run through them so that the cars can stop and be loaded or unloaded inside the operation. A seismographic analysis may be needed to demonstrate this, as shown next.

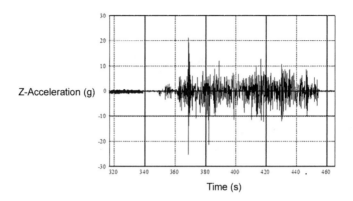

Facility Infrastructure

After you have documented the environmental conditions, you will need to examine the infrastructure of the facility or space where your system will be deployed. First, you should try to get a copy of the building plans or blueprints. Other useful documents can be network cabling diagrams, HVAC system diagrams, electrical diagrams, and so on.

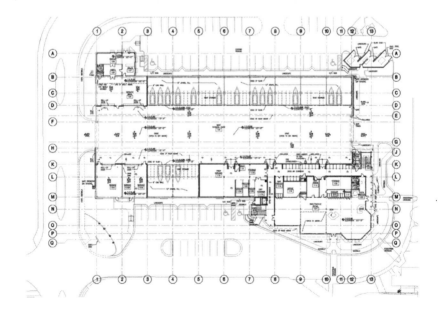

Sometimes these diagrams may not be available or accessible; you can use any facility plan or drawing that was originally designed for evacuation purposes instead. Once you get the building plans, you can start to mark information on the plans to correlate the facts that will drive decisions about where your IZs will be located and how they will look.

Location and Movement of Equipment and People

You should document all equipment and machinery used, including the models and possibly descriptions if they are customized for that particular building or operation. Examine equipment such as fork trucks, pallet jacks, and other vehicles that will be traveling through or near the interrogation zones. You must capture how vehicles, equipment, and personnel move about the location as part of the normal business process/operation. You should note the paths for vehicles, products, and personnel in your building plan. This will help you determine the placement of IZs and their protection, and it will help you avoid blocking paths and access/exit routes for vehicles and personnel.

Building Details

You must take notes about the structure and size of the building and what materials are used in the construction of the floors, walls, and roof. Measure the distances between dock doors and accurately mark the location of stationary objects that you will not be able to remove to accommodate IZs. You should also identify the location of the back-end processing computer that will be sending and receiving data to and from your RFID system. You must determine how data will move to and from that computer, which means identifying cabling and network requirements.

Travel Advisory

Let's talk a bit about cable length. Being 400 feet from the nearest wiring closet may not mean anything to you, but a professional network installer (or a person who has installed a network a couple of times) will know that the limit for copper wire is 328 feet. Therefore, you can't run a network wire that far and get your signal to and from the reader without a repeater in line. So, if the reader needs to be 400 feet from the wiring closet, plan to spend another $100 or so in your budget for a repeater. If you don't do this, your IZ might not be able to communicate with your back-end system.

If you're working on a multistory implementation, find out what is above and below the site where you are installing the system, and determine whether RFID systems on multiple floors need to be considered. You need to watch for potential sources of interference—consider whether you are installing an RFID system for asset management or a 2.4 GHz active real time location system (RTLS). On the floor below, for example, may be a company that manufactures wireless networking equipment, which happens to operate at 2.4 GHz, and this could cause interference. Look for anything that might affect the frequency you select for your system. You can acquire RTLS systems that run at other frequencies. Or it might be appropriate to select a different approach to avoid a potential source of interference if you can't remove it altogether. Equipment such as large machines and large motors can cause interference with RFID systems as well. You should know where that equipment is located relative to your IZs and where and when you can expect interference. (For more on this, see "RF Characteristics.")

Document the location of entry and exit doors, dock doors, and any "pinch points" for material moving throughout the facility. Note locations of the power outlets, which will help you determine whether new power lines are required or you can take advantage of existing power outlets.

Also note the location of access points, hubs, routers, switches, bridges, and any network equipment for LAN or wireless networks, including the coverage area of those networks and their security protocols. When you are selecting equipment, you should try to match the networking characteristics. If you are implementing equipment that will use an existing wireless network, make sure that it meets the security standards so that it functions properly on that network.

Mark the locations of stationary heavy machinery and conveyor belts, and mark the locations and dimensions of any restricted zones around machines. Document the policies about working or not working with a particular machine or device while it is running. In addition, look at equipment emissions, such as heat, steam, or any leaking chemicals that may affect the environment and your RFID equipment. Many machines, as well as heavy-duty cabinets and shelving, are made of metal, which can cause reflections or block the RF waves altogether.

Document materials in which the building, equipment, machinery, and products are made. This is especially important for materials that are not RF-friendly. RF friendliness will depend on proposed frequency of your system. (For more on RF-friendly materials, see Chapter 2.)

Planning the IZs

When you are performing a site analysis, you are also trying to determine where to place IZs, interrogators, printers, and antennas. You must also consider the flow of the goods; the antennas must be positioned near the goods to be tracked, yet out of harm's way. Knowing that when you are reading tags on assets, products, or people as they go through the zone, you want the antennas facing the zone, facing those products, facing those people, but still out of harm's way as much as possible.

If the facility has a large dock door, vehicles may carry tall pallets stacks, and a tall IZ with perhaps additional antennas will be required to cover the entire space. If slow-moving objects are being tracked, you could limit the number of antennas used, especially if you know that the tags are always going to be on one side, and the pallets always move the same way through the IZ (on a pallet conveyor, for example). For low-speed applications, only a couple of antennas may be required to get the job done as opposed to high-speed applications that need multiple antennas along the travel path to increase tag dwell time.

Objective 9.2 **RF Characteristics**

Conducting an RF site analysis in the facility may be required depending on the overall scope of the implementation. If you are implementing a simple "slap and ship" operation, you will not usually need to perform an RF site survey. However, if you are performing a multi-dock door and conveyor integration, an analysis of the environment from the RF perspective may save time and money because you will understand potential for interference. The analysis will also help you devise a strategy on how to avoid or overcome this interference.

If you are not experienced in performing RF site surveys, you should hire a specialized company to conduct the analysis, such as a local wireless network company. You do not need to survey the whole facility; you may target just the areas of IZs and potential interference.

In a full-blown implementation, sources of interference need to be identified with a spectrum analyzer, which will measure across a broad frequency range looking for what RF activities are already present at the site. Typically, you should examine a relatively wide band around the frequency you will be using for your system. If, for example, you are performing a site survey specific to 915 MHz, you will set your spectrum analyzer from 400 MHz to 1.5 GHz to measure all RF activity in the range.

You should also look for *harmonic* signals, which can cause problems with your system. In a European installation with a 433 MHz active tag RTLS and a passive system running between 865 and 868 MHz, for example, you should be aware that these signals could cause problems. 866 MHz is exactly double the 433 MHz value and is therefore the first-order harmonic, which could cause interference. Knowing this can save you a lot of headaches should problems arise in the future.

Local Lingo

Harmonics *Harmonics* occur when two wavelengths intersect. If one wavelength is exactly double the length of another, they can intersect and cause a null within the read zone.

How to Use a Spectrum Analyzer

Once a potential frequency interferer has been identified on the spectrum analyzer, it's pretty simple to identify the source of that frequency. Move 10 feet in any direction and note whether the signal gets stronger or weaker. If it gets stronger, continue to move in that same direction until the signal is at its strongest; then look in the immediate proximity for the source of that interference.

Dealing with Interference

If you find a potential source of interference, you will have to find a way to shield against it or you may have to replace it. If the problem is caused by two-way radios, for example, the solution may be easy: replace them with communication devices operating in a different frequency range. On the other hand, if the interference is caused by a manufacturing execution system, such as a crane and that can cost a couple million dollars, replacement may not be an option. For situations in which you cannot avoid interference for other reasons, you can consider shielding.

Some sources of interference can be conveyor belts, large motors running at 480 volts, electric motors on forklifts, electric power jacks, any kind of heavy machinery, and air-conditioning units. All of these devices can cause substantial interference to your low frequency (LF) and high frequency (HF) systems. Other objects may prove to be unexpected sources of interference. In one particular facility, for example, forklift operators honked their horns when they approached an intersection. When an operator honked the horn, it created a harmonic exactly in the 915 MHz band, causing interference with the UHF RFID system. The solution in this case was to prohibit use of horns in that immediate area or replace the horns with others that generated higher or lower tones.

Travel Advisory

Don't forget about harmonics created by active 433 MHz systems, which can interfere with UHF systems in Europe, which operate at 865 to 868 MHz.

When you are performing a site survey, ask the IT staff questions regarding any devices that could possibly cause interference: What types of two-way communication devices are employed here? What types of pagers are used? If a particular carrier is used for these devices, get the carrier name and a phone number to find out more information about the system if needed.

Many types of sudden interference are caused by faulty devices that should be fixed or replaced. Other sources of interference can be the following:

- Wireless network (usually operating at 2.45 GHz)
- Short-range radios
- Cordless phones (operating at 915 MHz or 2.45 GHz)
- Microwave ovens
- LAN equipment
- Neon lights
- Public or personnel address systems

Travel Assistance

Here's a story from a real-world implementation that may inspire you when solving hard-to-identify interference. In a 915 MHz installation, it was difficult to identify an interfering device, because the signal would often disappear and appear independently on running equipment. After a while, an engineer using a spectrum analyzer measured a large spike in a signal when someone was being paged through the internal speaker system. This led to the discovery of the problem: a speaker with a malfunctioning coil created signals in a frequency that caused a direct interference with the installed RFID system.

RF Path Loss Contour Analysis

If you run into a problem during installation or troubleshooting, it may be necessary to generate a path loss or RF path loss contour analysis (PLC analysis). You'll know that it may be beneficial to perform the RF PLC analysis, for example, when you have installed some equipment, and all the indicators say it works fine but you notice sporadic sources of interference that you cannot identify or spots where the tags are not read.

The PLC analysis measures RF signals all the way around your active IZ and provides information about the signal strength in granular detail. You'll need to do the following to perform this type of analysis:

1. Grab a spectrum analyzer and a measuring tape.

2. Measure signals in 1-foot increments from the antenna in all dimensions (straight, right, left, up, and down).

3. Mark the signal presence and, ideally, its strength in all points.

4. You can then generate a map and color code the areas according to signal strength. You can do this manually or using specialized software.

After you are done with the PLC analysis, you can determine how your signal propagates and identify where issues in the IZ exist, and you will also discover the sources of interference that you did not expect. Typically, PLC analysis is very expensive, as it is a time and labor-intensive process; however, it may pay off if your system is not performing as it should and you cannot determine the cause. An example of an RF PLC is shown here:

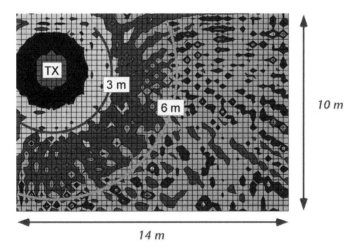

Difference Between Wi-Fi and RFID Site Analysis

RFID site assessments are not that different from wireless network implementation site surveys. However, Wi-Fi tries to achieve a result quite different from that of the RFID implementation. The goal of Wi-Fi implementation is to flood the entire facility with RF, while the goal of RFID is to flood only a particular area with a high likelihood of reading tags as they traverse that area (IZ) and control/contain the RF signals within this zone.

Objective 9.3 Business Process Documentation

An essential part of a site analysis is the business process analysis. To design an RFID system that will track your material, goods, assets, or personnel successfully and efficiently, you have to be familiar with the current business process, how the RFID technology can be integrated into it, or how you will have to change the business process to achieve the best results.

What to Document

To document all business processes at the site, you need to capture information such as how and when material moves, how equipment such as forklifts and cranes move, how and where raw materials are used during the process and how they are moved throughout the facility, and what happens with finished products. For instance, if this is a manufacturing line, is the product staged there at the facility for shipment or is it shipped in real time as it is manufactured? Or perhaps the products are manufactured and stored in a temporary warehouse attached to the manufacturing facility and are shipped after the order is complete. All of these processes depend on the type of facility and its operations.

You should also document all data about production lines and the machines used in them. The same machines can be used to manufacture multiple types of products. In such a scenario, a machine may manufacture a particular product for two weeks, and then the following week the machine is turned off and reconfigured to be used for manufacturing a new product. You may need to know about the production schedule for particular machines because different types of products will interact with your RFID system differently. Knowing all this ahead of time will help you plan a more efficient system.

Points of Contact

When you arrive on site, you will interact with a few important people, such as the facility manager (building manager or a plant manager), a safety manager, shift supervisors, and workers themselves. Why should you talk to these people? Because you do not want to be an uninvited intrusion into their worlds. You also need access to their knowledge of the site business processes to help you understand the way things work at the site.

The plant or building manager can easily differentiate between smooth procedure operation and any operation that is malfunctioning. The facility manager knows the site well, and although he/she probably won't know the granular details, this person can help you understand about the business process and facility policies. When you talk to building or plant managers, get information on the size of the building, the environment of the building, work schedule, work flow, and cycles of the workers. The general managers will know the schedule intimately and will be able to tell you when is the best time to work or which groups are going to be most receptive and least problematic. This information will hopefully help you accomplish your objectives while installing the RFID equipment.

You should also discuss the expected duration of the various phases of the project. Be realistic, and add a couple of days to the estimate so that you don't create false expectations in case some difficulties are encountered. With a facility manager, negotiate to determine the appropriate date and time for site analysis and installation. This should happen preferably during the low cycles of the workday.

Inform management about how the site analysis and system deployment will affect the facility operation. For example, tell them that, during the system installation, certain portions of the warehouse may be obstructed or filled with equipment and installers.

You can also ask about equipment they can provide, such as a scissor lift to access the ceiling, power tools, and perhaps specialized personnel such as facility electricians, IT personnel, and others who could help you in your installation and know the facility well. This conversation will help you determine the tools and resources you can access on site, who you have to hire, and what you will have to rent or ship from your home site. Using the facility resources is a lot less expensive not only for you, but also for your customers, since they will ultimately end up paying for any rental or shipping costs you encounter while installing the system.

The shift supervisors can tell you who are your key players. When you start deploying the RFID solutions, you'll need to make sure that you have access to a worker or a supervisor to assist you with deployment and/or troubleshooting, and you should familiarize this person or persons with the equipment sufficiently so that they can help you resolve any minor issues around the RFID system. If you can train key employees who are well respected among their peers, you become the second-level support, limiting the amount of calls you get at 2 o'clock in the morning: "Hey! This is not working! What do I do?" Instead of them calling you, they can ask a knowledgeable coworker trained by you for help in solving many problems. A shift supervisor must approve the engagement of such personnel in your endeavors.

You may also encounter negativity from workers at the facility regarding the RFID system's installation. For example, some may believe that the implementation is automating a task that was formerly performed by a human, thus the system is taking work away from employees. To prevent this perception and negative feelings, work on a communication plan with the customer's human resources personnel. You can also offer to conduct one-on-one or group discussions regarding the technical implications of the RFID implementation. Management should address the business reasons behind selecting and implementing an RFID system, what it means for employees, and how the new system will help the company as well as them. A well-designed system will add to the efficiency of the overall operation—whether that operation is manufacturing, warehousing, or distribution—and you should make sure that everybody knows this. Onsite management should explain to employees how the system will benefit them in their daily work. From time-to-time, a new system may result in staff reductions. Explanation and justification of this to employees is the job of company management.

That said, don't be afraid to talk to workers and other personnel. Ask questions, such as "Do you think this will work for you?" or "Is there anything I can do to make this better than what I am proposing now?" If you get the workers involved from the beginning, they will feel included in the process and may offer good pointers that will help your system design.

Objective 9.4 Site Survey Report

W hen conducting a site survey, you are also preparing to generate a site survey report. Ask questions, take pictures, and record everything you find out. This information is necessary in creating a thorough report.

During a site survey, the person doing the site analysis will not always be the same person who is designing the overall solution. Instead, a team of people will contribute to the design of the system. Photos of the proposed IZs will make it easier for you to explain to the team all site challenges as well as solutions. Team members may also notice things you missed while they examine the photos. You can draw on their experience as well as your own.

Regulations and Company Policies

You must follow local regulations as they apply to RF emissions and equipment installation. For example, a certified electrician must install system wiring, and this person will be aware of local regulations regarding electrical work. Professionals will make sure that the electrical wires are installed in conduit and that all equipment is properly grounded so that electrical devices operate safely. Local regulations may also require that the network cables be installed in a steel conduit. Such requirements vary from county to county, state to state, and country to country.

You may also have to deal with union regulations. In some shops, only union installers are allowed to install network, electrical, or electromechanical devices. Unions also require strict safety procedures. Therefore, you must determine who in the company can do what work at what time, and what type of external labor must be used during the installation. Here's a real-world example to illustrate this point. While we were installing a forklift solution at a paper mill, I noticed that the motor shop was overrun with repairs that day. I began installing RFID equipment to help out. A union employee saw me do this and reported me to the union. The next day, I was called into the mill manager's office to explain why I was turning wrenches on their trucks!

Company regulations and policies are also important to understand, because your noncompliance may cause big delays and installation problems. For example, you may need entering and exiting permits for certain equipment and materials required for your implementation. Other company regulations and policies may require personal protective equipment such as steel-toe boots, hard hats, or safety guards in order to step into certain areas or operate certain machinery. You also must have a special license to operate a lift, fork truck, or electrical pallet jack. It's important that you know about licenses, certifications, and insurance coverage you need before you begin your work.

You may be required to be accompanied by an employee any time you work in a particular area due to safety or security reasons, or you may need security clearance (such as on military sites) or even drug testing to participate in activities at a particular worksite.

Travel Assistance

Make sure you ask about company policy regarding using cameras in a facility. Some companies are strict about camera usage and require that you get permission before taking pictures. You may also be required to provide photos for approval before you leave the site.

CHECKPOINT

✔**Objective 9.1: Infrastructure Assessment**   Take notes about the structure and size of the building and the materials used in the construction of the floors, walls, and roof. Capture the building layout, including static equipment and paths of moving equipment and personnel. Take measurements of all spaces where future IZs may be located.

When analyzing the environment at the site, record the temperature and humidity variations. Document the type of facility in which the system will be deployed and accordingly note conditions such as humidity experienced in the facility, any chemicals that are used in the process, presence of industrial dust and dirt, the presence of high-pressure washing, and other such information. Equipment can be subjected to vibration not only when installed on or near vehicles such as forklifts, but also on conveyors, manufacturing lines, and heavy machinery inside the facility; rail lines, highways, or airports outside of the facility but in the near proximity can also affect vibration.

✔**Objective 9.2: RF Characteristics**   Sources of interference can be identified by checking operating frequencies of equipment used within the site and using a spectrum analyzer. Interference can be caused by wireless networks (usually operating at 2.45 GHz, but older types use around 900 MHz), short range radios, cordless phones (operating at 915 MHz or 2.45 GHz), cell phones (operating at 900 MHz), microwave ovens, LAN equipment, neon lights, public or personnel address systems, conveyor belts, large motors running at 480 volts, electric motors on forklifts, electric power jacks, or any kind of heavy machinery as well as air-conditioning units. Also look for harmonic interference.

Set your spectrum analyzer for a slightly wider bandwidth than the frequency band you are investigating and watch for signal spikes. Move toward potential sources of these signals and note whether the signal gets stronger or weaker. The strongest signal will be the closest to the interfering device.

✔**Objective 9.3: Business Process Documentation**   An essential part of a site analysis is the business process analysis. To design an RFID system that will track materials, goods, assets, or personnel successfully and efficiently, you will need to know what the business process is now, how you integrate the RFID technology into it, or how you will have to change the business process to achieve the best results.

✔**Objective 9.4: Site Survey Report**   When performing a site survey or site assessment, you are also preparing to generate a site survey report. Be sure you always ask questions, take pictures (with permission if necessary), and record other important details in the site survey report, because this report will form the basis for your design of the site's RFID system.

REVIEW QUESTIONS

1. Why is it important to train key users of the RFID system?
    A. So that they can implement the RFID system themselves.
    B. It reduces the cost of hiring someone to do the troubleshooting.
    C. It provides base level support for basic problems.
    D. It makes the management happy.

2. During a site analysis for a UHF RFID system, which of the following documents can help the most in locating potential interference?

    A. Network cabling diagram
    B. Building maintenance records
    C. Production schedule
    D. 2.45 GHz wireless network diagram

3. You are asked to deploy an RFID system in two identical warehouses in different locations. You already performed a site survey including RF analysis in one of the warehouses. Do you need to perform it again in the second warehouse?

    A. No. The facilities are identical.
    B. No. You can troubleshoot the problems later.
    C. Yes. You want to get more billing hours.
    D. Yes. The facilities may look identical but they can have different workflow patterns as well as different RF characteristics.

4. Which of the following items should be documented in the site analysis report? (Choose two answers.)

    A. Use of wireless communications
    B. Paths of moving equipment
    C. Locations of bathrooms
    D. Average temperature within the site

5. Can you easily implement a 433 MHz active RTLS system and a passive UHF system within the same site in Europe?

    A. Yes. 433 MHz does not interfere with European frequencies.
    B. No. The European regulations forbid using RTLS systems.
    C. No, due to harmonic interference.
    D. No, due to multipath interference.

6. Which of the following statements applies when trying to identify the source of RF interference using a spectrum analyzer?

    A. The farther you move away from the source of interference, the stronger the signal shown by the spectrum analyzer.

B. The closer you move to the source of interference, the stronger the signal shown by the spectrum analyzer.

C. The source of interference cannot be detected using a spectrum analyzer.

D. The distance of a spectrum analyzer from the source of interference has no effect on signal strength.

7. You hired a consultant to perform RF site analysis for your facility, where you will be deploying a UHF RFID system. Which of the following equipment should he/she use?

   A. A spectrum analyzer with range of 100 MHz to 2 GHz

   B. An oscilloscope to measure RF wave oscillations

   C. A spectrum analyzer with a range of 100 kHz to 900 MHz

   D. Linearly polarized Gen 2 UHF interrogator and passive transponder

8. When performing a site survey, which of the following documents is useful?

   A. An RFID equipment price list

   B. An organization chart

   C. A facility blueprint

   D. None of the above

9. Which of the following is not a part of a site survey?

   A. A workflow documentation

   B. Maintenance history of machinery

   C. Electrical wiring layout

   D. Network layout

10. When performing a site survey, you should take photos. What are the rules?

    A. Make sure that no one in the facility sees you taking photos.

    B. Always acquire a permit from facility management.

    C. Ask the workers for permission to take photos.

    D. As long as there are no name brands on the photos, you do not need to ask for permission to shoot.

REVIEW ANSWERS

1. **C**   It is important to train key users of the RFID system so that you can get a base level support at the site, in which trained users can perform some basic troubleshooting tasks or be able to perform them with a telephonic supervision.

2. **C**   During a site analysis for a UHF RFID system, the most helpful documents in locating potential interference will be the production schedule.

3. **D**   Although the warehouses may look identical, they can have different workflow patterns as well as different RF characteristics. Therefore, you should perform the RF site analysis on the other warehouse.

4. **A** **B**   In the site analysis report, you should document the use of wireless communications as well as paths of moving equipment. You do not have to worry about the average temperature, unless the temperatures are extremely variable.

5. **C**   When implementing a 433 MHz active RTLS system and a passive UHF system within the same site in Europe, you should be concerned about the harmonic interference.

6. **B**   When trying to identify the source of RF interference using a spectrum analyzer, the closer you move to the source of interference, the stronger the signal shown by the spectrum analyzer.

7. **A**   To perform RF site analysis for a facility in which you will be deploying a UHF RFID system, the consultant will use a spectrum analyzer with range of 100 MHz to 2 GHz.

8. **C**   When performing a site survey, a facility blueprint is one of the most useful documents.

9. **B**   Maintenance history of machinery is not usually a part of a site survey.

10. **B**   When performing a site survey and taking photos, you must always acquire a permit from facility management.

Standards and Regulations

ETA	NEWBIE	SOME EXPERIENCE	EXPERT
	6 hours	3 hours	1 hour

This chapter discusses various standards and regulations related to the design and use of RFID systems. It also provides a brief summary of a few RFID mandates issued by some commercial and governmental organizations, which drive a large part of RFID implementations today.

Standards are created by various organizations to facilitate interoperability among components of the system designed and manufactured by many different organizations. Standards define not only hardware and software design, manufacturing, and development techniques, but they also define its use. Many organizations develop standards—for example, the International Organization for Standardization (ISO), American National Standards Institute (ANSI), and EPCglobal.

Regulations are developed by government regulatory agencies, such as Federal Communications Commission (FCC), Food and Drug Administration (FDA), and Federal Aviation Administration (FAA). While more than one *standard* may be available for a particular use and compliance with it is optional, *regulations* must be obeyed. *Mandate* compliance is optional. Mandates are created by organizations as a policy to interact with various business partners. The following sections describe various RFID standards, regulations, and mandates.

Objective 10.1 Standards

Many types of organizations create standards:

- Standards-creating organizations such as ISO, ANSI, and European Telecommunications Standards Institute (ETSI)
- Technology groups such as Institute of Electrical and Electronics Engineers (IEEE), American Society of Mechanical Engineers (ASME), and American Society of Civil Engineers (ASCE)
- Governmental agencies such as Department of Defense (DoD) and FCC
- Industry consortiums such as Automotive Industry Action Group (AIAG) and EPCglobal
- Commercial companies such as IBM, Microsoft, and Cisco

Standards are created to provide interoperability among different components of a system, interchangeability of components manufactured by different vendors, and a consistent user interface for similar systems created by different organizations. Some of the benefits provided by standards are reduced cost, availability of parts even if the original manufacturer has gone out of business, reduced user training, and ease of maintenance. Standards also help raise levels of quality, safety, reliability, efficiency, and interchangeability. Standards not only assist engineers and manufacturers by helping to solve basic problems in production and distribution, but they also benefit product users. Use of a standard is typically voluntary, and in some cases a manufacturer may even choose to comply with only one standard from a group of similar standards. When many similar standards exist—for example, standards that apply both to IBM PCs and Apple Macs—the industry becomes segmented and some of the benefits of standards are reduced. However, such variations also provide paths to alternative technologies.

Some of the disadvantages of standards are the slow acceptance of new technologies, the length of time required for standards creation, and the fact that standards are developed by a committee of people who have vested interests in standardizing the technology developed by their particular companies. To quote a popular adage, "Standards are like sausages: you don't want to know how they are made." To accommodate the interests of various factions within the committee, standards often do not implement the technology in an optimal manner. Small companies are slow in improving the existing technology because it may require the creation of new standards. But the advantages of standards far outweigh the disadvantages, which is evident by the existence of so many standards. That is why people jokingly say "The good thing about standards is that there are too many to choose from."

During the initial phase of new technology development and implementation, such as RFID, no standards exist, or they are created by a few companies that are pioneers in the new technology. The standards created by a company are the intellectual property of this company, and other manufacturers may have to pay royalties to use them. These types of standards are called *proprietary* standards. Standards created by standardization organizations are open and are available for free or for a small fee that is paid by all the users of the standard. Until around the middle of 2005, no globally accepted open standards existed for air interface between the RFID tag and the interrogator. While the open standards were being developed, growth of the RFID technology was in a holding pattern, and many manufacturers and users were sitting on the fence, waiting. With the emergence of the new ISO 18000-6C standard in mid-2006, discussed later in the chapter, the RFID technology has moved into the fast lane. More standards are needed for collecting and exchanging RFID data. These standards are in various stages of development, and as products complying with those standards become available, the use of RFID technology may explode.

Overall, standards provide many benefits to businesses, governments, and consumers. Businesses benefit because the widespread adoption of standards means that suppliers can base the development of their products and services on specifications that have wide acceptance in their sectors. This, in turn, means that businesses using standards can freely compete in markets around the world. Customers get wide choices of products due to the worldwide compatibility of technology achieved by use of standards, and they also benefit from the effects of competition among suppliers. For governments, standards provide the technological and scientific bases for health, safety, and environmental legislation. Consumers enjoy the benefits of conformity of products and services with assurances about their quality, safety, and reliability.

Standards Organizations

As mentioned, many organizations develop standards. They may be international, regional, national, or industry consortium. Following are some of the standards-creating organizations:

- International
  - ISO: International Organization for Standardization
  - IEC: International Electrotechnical Commission
  - ITU: International Telecommunication Union
- Regional
  - CEN: European Committee for Normalization and Standardization
- National
  - ANSI: American National Standards Institute
  - BSI: British Standards Institution
  - SCC: Standards Council of Canada
  - AFNOR: Association Française de Normalization (the French industrial standards authority)
  - JISC: Japan Industrial Standards Committee
- Industry (these may be international)
  - AIAG: Automotive Industry Action Group
  - EPCglobal
  - UCC/EAN: Uniform Code Council/European Article Numbering System

Various standards-creating organizations have different detailed development processes, but the general process is similar. A working group generates a document that describes the standard in detail. This document is created after several technical discussions among members of the group who are experts in the technology. It is then circulated to the members of groups involved in similar technologies and their suggested changes are considered. Then the document is opened to the general public for comments. Once the issues raised are satisfactorily resolved, a proposed standard is submitted to various committees for approval. When that process is completed, it becomes a ratified standard. The entire process may take from one to three years. The standards-developing organizations discussed in this chapter are ISO, IEC, ETSI, European Radiocommunications Office (ERO), DoD, and EPCglobal.

ISO and IEC

The ISO is a network of the national standards institutes of 157 countries. Each member country, no matter how large or small, gets one vote. The central secretariat of ISO, located in Geneva, Switzerland, coordinates the entire organization. ISO is a nongovernmental organization and its members are not, as is the case in the United Nations system, delegations of national governments. The participating country selects a domestic organization to be its representative to ISO—for example, ANSI for the United States, SCC for Canada, AFNOR for France, BSI for Great Britain, and JISC for Japan. Many of its member institutes are part of the governmental structure of their countries, while others are set up by national partnerships of industry associations. ISO standards meet the requirements of businesses and the broader needs of consumers and users. Since its birth in 1947, ISO has published more than 16,000 international standards.

Just like any other standards, ISO standards are voluntary. ISO has no legal authority to enforce its implementations. Many countries adopt ISO standards as part of their regulatory frameworks or refer to them in legislation as the technical basis. Many ISO standards, such as ISO 9000, have become a market requirement. ISO develops only those standards for which there is a market requirement. The work is carried out by experts from the industrial, technical, and business sectors that have asked for the standards and that subsequently put them to use. Of the approximately 3000 ISO technical groups (technical committees, subcommittees, working groups, and so on), about 50,000 experts participate annually in development of ISO standards.

The IEC is the leading global organization that develops and publishes international standards for all electrical, electronic, and related technologies. These serve as a basis for national standardization and as references when drafting international tenders and contracts. IEC was created in 1906 and currently has more than 130 participant countries, including 67 members and 69 affiliates. Its headquarters are in Geneva. The IEC charter includes all electrotechnologies, including electronics, magnetics and electromagnetics, electroacoustics, multimedia, telecommunication, and energy production and distribution. The work is done by 10,000 electrical and electronics experts from industry, government, academia, and test labs. IEC standards use numbers in the range 60000–79999. IEC also develops standards jointly with ISO (as you can see in Figure 10-1). Such standards are designated by ISO/IEC followed by a number. Only the standards developed by ISO/IEC Joint Technical Committee 1 on Information Technology use the ISO/IEC prefix.

Figure 10-1 shows Automatic Identification and Data Capture (AIDC) focused ISO committees and their interrelationships.

ERO

The European Radiocommunications Office provides a center of expertise that acts as a focal point, identifying problem areas and new possibilities in the radio and telecommunications fields. It supports and advises the Electronic Communications Committee (ECC). ERO develops long-term plans for future use of the radio frequency spectrum in Europe. It also supports and works together with national frequency-management authorities in European countries. In addition, it identifies and promotes best practice in administration of national numbering schemes and numbers. ERO is the distribution point for all ECC documents and also provides detailed information about the work of the ECC via the ERO Web site, www.ero.dk. The ERO site provides information about the latest developments within the ECC with reports of recent meetings and ap-

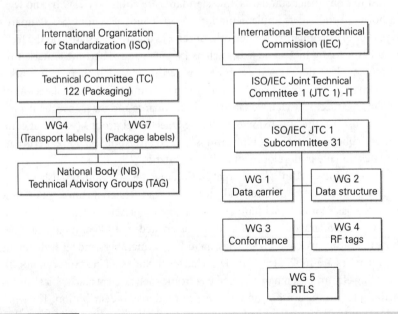

FIGURE 10.1   ISO and IEC committees for AIDC

proved texts of ECC decisions, recommendations, and reports. Many organizations such as government departments, public radiocommunications operators, manufacturers, users, private network operators, service providers, research institutes, and standards-making bodies participate in ERO work. ERO has the overall objective of developing proposals for a European Table of Frequency Allocations and Utilizations for the frequency range 9 kHz to 275 GHz.

CEN

The European Committee for Standardization was founded in 1961 by the national standards bodies in the European Economic Community and European Free Trade Association (EFTA) countries. CEN contributes to the objectives of the European Union and European Economic Area with voluntary technical standards that promote free trade, worker and consumer safety, network interoperability, environmental protection, research and development programs, and public procurement.

EPCglobal

EPCglobal is a neutral, consensus-based, not-for-profit standards organization that functions as a consortium of technology companies and the user community. It is a joint venture between EAN International and the Uniform Code Council (UCC), now called GS1. It was created in October 2003 from the Auto-ID Center at Massachusetts Institute of Technology (MIT). The Auto-ID Center was created by UCC, Procter & Gamble, and Gillette in 1999. Later, more than 100 companies joined to develop EPCglobal Network. In 2003, when EPCglobal was created from the Auto-ID Center, the research work done by the center was transferred to Auto-ID Labs created at six universities in different countries. In 2004, it created a standard for the air interface protocol at the 860–960 MHz frequency called Generation 2, or Gen 2. It was submitted to ISO for ratification in early 2005, and it was approved and became an ISO 18000-6C standard in July 2006.

EPCglobal is now developing standards for the EPCglobal Network, a network to exchange data collected by RFID devices among various partners in the supply chain. It is working toward establishing the EPC Network as the global standard. The goal of the EPC Network is to create worldwide real-time visibility of objects in the supply chain. EPCglobal functions are conducting research and development work on RFID (in collaboration with Auto-ID Labs); creating standards and regulations for tags, interrogators, and information systems; managing the EPCglobal Network including its numbering system, Object Name Service (ONS) and certification services; and driving adoption via marketing, awareness, business development, and implementation support.

DoD

The US Department of Defense is a large (probably the largest in the world) consumer of services and goods. It has a budget of well over $500 billion and 60,000 suppliers. To make products from different suppliers interoperate, it creates its own standards called Military Standards or Mil-standards, available at the DoD Web site. The standards must be met by all goods supplied to the DoD.

US Defense Logistics Agency (DLA) Automatic Identification Technology (AIT) Office is the foundation in the DoD's efforts to provide timely asset visibility in the logistics pipeline, whether in-process, in-storage, or in-transit. AIT media includes barcodes, radio frequency ID, satellite tracking systems, smart cards/common access cards (CACs), optical memory cards, and contact memory buttons. By enabling data collection and transmission to automated information systems (AISs), AIT provides the various military commanders with the ability to track, document, and control the deployment of personnel and material.

UCC

The mission of the Uniform Code Council, Inc., is to take a global leadership role in establishing and promoting multi-industry standards for product identification and related electronic communication. The goal is to enhance supply chain management, thus contributing added value to the customer.

EAN International

European Article Numbering International is designed to create global open standards based on best business practices and to drive their implementation. Its goal is to play a leading role in improving supply and demand chain management worldwide.

Types of Standards

Four categories of standards are typically used in RFID systems:

- Technology standards
  - Define how various hardware and software should be designed. These standards provide the details of communications between the interrogator and tags, the modulation of analog signals, coding schemes for digital data, and the interrogator commands and tag responses.
  - Define the interrogator interface with the host.
  - Define the data syntax, structure, and content.
- Data content standards
  - Define the meaning of bit streams read from the RFID tags and provide guidelines on how the data shall be presented to applications.

- Specify the commands that are supported for transferring data between the application and the tag.
- Provide details of data identifier, application identifier, and data syntax.
- Conformance standards
  - Define test methods for determining the conformance of devices (tags and interrogators) to a specific standard.
- Application standards
  - Define how the technology is implemented for a particular application. For example, standards for the freight container identification system will define where and how the RFID tag should attached to the container.
  - Provide details on labeling, product package, and numbering schemes.

Following are some examples of standards used in RFID systems:

- Technology standards
  - **ISO 18000**   Defines air interface standards between interrogator and tags at various frequencies
  - **EPC Gen 2**   Defines air interface standards at 860–960 MHz frequency
- Data content standards
  - **ISO/IEC 15424**   Data Carrier/Symbology Identifiers
  - **ISO/IEC 15418**   Application Identifiers & Data Identifiers
  - **ISO/IEC 15434**   Syntax for High Capacity ADC Media
  - **ISO/IEC 15459**   Transport License Plate
  - **ISO/IEC 24721**   Unique Identification
  - **ISO/IEC 15961**   Data Protocol: Application Interface
  - **ISO/IEC 15962**   Data Protocol: Data Encoding Rules and Logical Memory Functions
  - **ISO/IEC 15963**   Unique ID of RF Tag
  - **EPC Tag Data Standards Version 1.3**
- Conformance standards
  - **ISO/IEC 18046**   RFID device performance test methods
  - **ISO/IEC 18047**   RFID device conformance test methods
    - Part 2: 125–150 MHz
    - Part 3: 13.56 MHz
    - Part 4: 2450 MHz
    - Part 6: 860–960 MHz
    - Part 7: 433.92 MHz (active)

- Application standards
  - **ISO 10374** Freight Containers: Automatic Identification
  - **ISO 18185** Freight Containers: Radio-Frequency Communication Protocol for Electronic Seal
  - **ISO 11784** Radio-Frequency Identification of Animals: Code Structure
  - **ISO 11785** Radio-Frequency Identification of Animals: Technical Concept
  - **ISO 14223-1** Radio-Frequency Identification of Animals: Advanced Transponders Part 1: Air Interface
  - **ISO 21007-1** Gas Cylinders: Identification and Marking Using Radio Frequency Identification Technology Part 1: Reference Architecture and Terminology
  - **ISO 21007-2** Gas Cylinders: Identification and Marking Using Radio Frequency Identification Technology Part 2: Numbering Schemes for Radio Frequency
  - **ANSI MH10.8.4** RFID for Returnable Containers
  - **AIAG B-11** Tire & Wheel Identification Standard
  - **ISO 122/104 JWG** Supply Chain Applications of RFID

RFID Standards

This section covers some of the important standards that are part of the CompTIA RFID+ Certification Exam.

ISO/IEC 18000

The ISO/IEC 18000 series of standards has seven parts and deals only with the air interface protocol. These standards are not concerned with data content or the physical implementation of the tags and interrogators. The standards define use of five frequency bands for communication between interrogator and tags. Five different frequency bands are used because RF waves of different frequencies are reflected, refracted, and absorbed differently by different materials; a tag antenna type and size varies with frequency; and maximum read range varies with frequency and legacy frequency allocations.

Exam Tip
Remember the frequencies related to different parts of ISO/IEC 18000.

ISO/IEC 18000-1   This is Part 1 of the 18000 series standards. It provides a framework for defining common communications protocols for globally useable frequencies for RFID. It also establishes generic parameters that could be determined in any standardized air interface definition in the ISO 18000 series. The specific values for air interface definition parameters related to any frequency are provided by the subsequent parts dealing with that frequency. This standard establishes a common system management and control and information exchange framework that can be used at various frequencies.

ISO/IEC 18000-2   Part 2 specifies parameters for air interface communications between the interrogator and the tag below 135 kHz frequency. This part defines protocol, commands, and methods for detecting and communicating with one tag among several tags (anti-collision), but the implementation of anti-collision is optional.

Two types of tags, Type-A and Type-B, are defined in this standard. They differ at the physical layer but support the same anti-collision and protocol. Type-A tags operate in full duplex (FDX) mode at 125 kHz frequency. They use the same channel for two-way transmission between tag and interrogator. FDX tags are permanently powered by the interrogator, including during the tag-to-interrogator transmission. Type-B tags operate in half duplex (HDX) mode at 134.2 kHz frequency. They use two different one-way channels for interrogator to tag and tag to interrogator transmission. HDX tags are powered by the interrogator, except during the tag-to-interrogator transmission.

To claim compliance with this standard, a tag must be of either Type-A or Type-B. In addition, an interrogator must support both Types A and B tags. Depending on the application, an interrogator may be configured as Type-A only, Type-B only, or as Type-A and Type-B. When configured in Type-A and Type-B, and when in the inventory phase, the interrogator shall alternate between Type-A and Type-B interrogation.

ISO/IEC 18000-3   This standard provides parameters for air interface communications at the 13.56 MHz frequency. It defines the physical layer, collision management system, and protocol values for RFID systems for item identification operating at 13.56 MHz frequency in accordance with the requirements of ISO 18000-1. This standard has two non-interfering, non-interoperable modes of operation, intended to address different applications. Both of the modes require a license from the owner of the intellectual property, which shall be available on terms in accordance with ISO policy. The interrogator must support either Mode-1 or Mode-2, or it could optionally support both.

Mode-1 is based on ISO 15693 with modification for item management and is intended to improve the compatibility between vendors. Mode-1 provides the following parameters:

- Interrogator to tag data rate: 1.65 Kbps or 26.48 Kbps
- Tag to interrogator data rate: 26.48 Kbps

Mode-2 allows for higher speed and memory (PJM, or Phase-Jitter Modulation) than Mode-1 and has the following parameters:

- Interrogator to tag data rate: 423.75 Kbps
- Tag to interrogator data rate: 105.9375 Kbps on each of 8 channels

ISO/IEC 18000-4   This standard defines the communications protocol used in the air interface for RFID devices operating at the 2.45 GHz frequency used in item management applications. It also defines the forward and return link parameters for technical attributes such as operating frequency, channel bandwidth, maximum power, modulation, duty cycle, data coding, bit rate, frequency hop rate, hop sequence, spreading sequence, and chip rate.

This standard defines two modes. The first mode is for passive tags operating as an interrogator talks first (ITF), while the second mode is for battery-assisted tags operating as a tag talks first (TTF). The detailed technical differences between the modes are provided in the parameter tables of the standard.

Mode-1 has the following characteristics:

- Applies in a passive backscatter RFID system
- System uses ITF technique
- One or more tags within the interrogation zone
- Range can be greater than 3 feet
- Much commercial activity for this mode

Mode-2 provides the following features:

- Used in long-range, high–data rate RFID systems
- Tag is battery assisted but operating as a TTF
- The air interface description does not list explicit claim for battery assistance in the tag
- Range greater than 300 feet
- Gross data rate up to 384 Kbps at the air interface in case of read/write (R/W) tag, while for read only (R/O) tag the data rate is 76.8 Kbps

ISO/IEC 18000-5   This standard defines the communications protocol used in the air interface for RFID devices operating at 5.8 GHz frequency used in item management applications. Due to the lack of commercial interest in the 5.8 GHz frequency RFID system, this standard was withdrawn.

ISO/IEC 18000-6   This standard describes the physical interactions between the interrogator and the tag, the protocols and the commands, and the collision arbitration schemes for passive RFID systems operating within the 860 to 960 MHz frequency range. It is intended to allow for compatibility and to encourage interoperability of products for the growing RFID market in the international marketplace. It defines the forward and return link parameters for technical attributes such as operating frequency, operating channel accuracy, channel bandwidth, maximum power, modulation, duty cycle, data coding, bit rate, and bit transmission order. The standard describes three noncompatible Types A, B, and C.

Type-A has these features:

- Uses pulse interval encoding (PIE) in the forward link
- Reader talks first
- Adaptive ALOHA collision arbitration algorithm
- Uses biphase space FM0 return link encoding
- Data rate 33/40 Kbps
- Frequency range 860–930 MHz

Type-B has these features:

- Uses biphase modulation and Manchester encoding in the forward link
- Reader talks first
- Adaptive binary tree collision arbitration algorithm
- Uses biphase space FM0 return link encoding
- Data rate 8/40 Kbps
- Frequency range 860–930 MHz

Type-C has these features:

- Equivalent to EPC Class 1 Gen 2 standard
- Discussed in detail in the section "EPCglobal Gen 2"

ISO/IEC 18000-7   Part 7 defines the air interface for RFID devices operating as an active RF tag in the 433 MHz band used in item management applications. Its purpose is to provide a common technical specification for RFID devices that may be used by ISO committees developing RFID application standards. This standard is intended to allow for compatibility and to encourage interoperability of products in the international marketplace. It defines the forward and return link parameters for technical attributes such as operating frequency, operating channel accuracy, occupied channel bandwidth, maximum power, spurious emissions, modulation, duty cycle, and data coding. Typical applications operate at ranges greater than 1 meter. It was developed for FCC-approved read/write active tags. These tags are used by US DoD and Universal Postal Union and have a read range in excess of 300 feet.

ISO 14443

This standard defines identification cards operating at the 13.56 MHz frequency using near-field inductive coupling. The cards are usually called *proximity* cards. Typical applications include identity, security, payment, mass-transit, and access control. The interrogators for these cards use an embedded microcontroller and a magnetic loop antenna that operates at 13.56 MHz frequency. More recent International Civil Aviation Organization (ICAO) standards for machine-readable travel documents specify a cryptographically-signed file format and authentication protocol for storing biometric features such as photos of the face, fingerprints, and iris. ISO 14443 systems are designed for a range of about 10 centimeters (3.94 inches), so they are a good fit for applications such as vending machines.

This standard describes two types of cards: Type-A and Type-B. The key differences between these types concern method of signal modulation, coding schemes, and protocol initialization procedures. Both Type-A and Type-B cards use the same high-level protocol.

Type-A is the most widely used contactless standard in the world, used mainly for transport applications. It is used by two products: Mifare, supplied by Phillips and Infineon, and version A of PicoPass, supplied by Inside Contactless. Type-B has a number of advantages over Type-A. It requires no patents on communication coding, is widely adopted by reader manufacturers with proven interoperability, and has been adopted as a national standard by countries such as Japan, China, and the United States.

ISO 15693

ISO 15693 is an ISO standard for vicinity cards, which can be read from a greater distance compared to proximity cards defined by ISO 14443. ISO 15693 systems operate at the 13.56 MHz frequency, use near-field inductive coupling, and offer maximum read distance of 3 to 5 feet. This range makes them a good fit for applications such as physical access or controlling entry to a parking garage, where it's inconvenient for users to open their doors or roll down a window just to get a contactless smart card close enough to the reader. It also serves as the foundation for a variety of applications outside of contactless smart cards, such as airline baggage tracking and supply chain management.

Communication from the reader to the card uses amplitude shift keying with 10 percent or 100 percent modulation index. The card sends its data back to the reader in two ways: amplitude shift keying and frequency shift keying. Both methods use the Manchester code to encode data. The data rate is 26 Kbps, which is about four times slower than that of 14443 standards but provides the increased communication range. The contactless cards usually transfer small amounts of data, so throughput isn't a deciding or noticeable factor. It has been widely adopted by access control applications and is currently penetrating other markets. Such an extensive adoption means more ISO 15693–compliant components with lower prices on the market as their volumes increase. Also, the design of the interrogator is simpler, giving further advantage in device cost.

Travel Advisory
Don't confuse ISO 15693 with ISO 15963, which is used for RFID for Item Management–Unique Identification of the RF tag.

ISO/IEC 15961

This standard establishes host-interrogator-tag functional commands and syntax features such as RFID tag types, data storage formats, and compaction types. The commands and syntax are independent of transmission media and air interface protocols. It is intended to be a companion standard to 15962, which provides the overall protocol for data handling. The 15961 standard comprises a super set of all functional commands and other syntax features appropriate to RFID for item management. The 15961 functional commands are at a higher abstract level than those of the ISO 18000 series. The interrogator-tag commands of 18000 series are at a detailed lower level and are specific to the particular technologies that are part of 18000.

ISO/IEC 15962

This standard establishes RFID for item management data syntax. It specifies the interface procedures used to exchange information in an RFID system for item management. Since no direct communication can occur between the host system and the RFID tag, the protocols established in this standard ensure the correct formatting of data, the structure of commands, and the processing of errors in the RFID system. This standard provides a basis of interoperability for current and legacy systems and a migration path to future systems.

ISO/IEC 15963

This standard establishes unique identification of RF tag and registration authority to manage the uniqueness. It specifies the numbering system for the identification of an RF tag, the registration procedure, and the use of it. The numbering system provides to the RFID application a means of uniquely identifying an RF tag. This number is encoded in the integrated circuit of the RFID tag. This standard consists of the following two parts:

- Part 1: Numbering system
- Part 2: Registration procedure and management guidance and rules

This standard permits addressing three main domains of the RFID system:

- Traceability of the integrated circuit itself for quality control in their manufacturing process
- Traceability of the RF tags during their manufacturing process and along their life in the applications in which they are used
- Anti-collision of multiple tags in the reader's field of view

EPCglobal Gen 2

This standard is officially called *EPC Radio-Frequency Identity Protocols Class 1 Generation 2 UHF RFID protocol for communications at 860–960 MHz Version 1.0.9*. It was developed by EPCglobal, Inc., and was approved as ISO 18000-6C in July 2006. It defines air interface parameters for tags operating within the frequency range of 860–960 MHz and allows for use of different frequencies in different regions from within this range. Different regions of the world have varying frequencies allocated for RFID in UHF range. For example, 902–928 MHz in the United States, 865–868 MHz in Europe, 908.5–914.0 MHz in Korea, and 952–954 MHz in Japan.

> **Local Lingo**
>
> **Gen 2**   You will see Gen 2, C1G2, Class 1 Generation 2, or ISO 18000-6C, but all of these terms refer to the previously-mentioned EPC Radio-Frequency Identity Protocols Class 1 Generation 2 UHF RFID protocol for communications at 860–960 MHz developed by EPCglobal and ratified by the ISO.

Key Gen 2 features are as follows:

- **Ability to change encoding according to the environment**   The reader changes the encoding method, the Miller sub-carrier or FM0, according to the noise in the environment. In a low-noise environment it may use FM0 encoding, which is faster, but as noise increases it may switch to

Miller sub-carrier, which is designed to optimize performance in noisy and dense reader environments. This decreases the number of tags read per second but allows tags to be read in harsher environments.

- **Three modes for reader operation**   The reader may operate in single, multi, and dense environments. A dense reader environment (see the following Local Lingo) is designed for enterprise deployments in which hundreds of readers are operating at the same time.
- **Tag population management**   Provides select, inventory, and access commands for efficient reading of tags—for example, a group of tags may be selected with a wild card pattern.
- **Longer kill and access passwords**   32-bits-long access and kill passwords increase the level of security for the data on the tags.
- **Forward link data protection**   Tags provide a randomly generated number to the reader to encode the data sent by the reader to the tags.
- **Four sessions for tag inventory**   Tag may operate in four different sessions at the same time, so four different readers can communicate with the tag at the same time without interfering with each other.
- **Faster data transmission rate up to 640 Kbps**   This is five times faster than the previous standards.
- **Improved tag memory and programmability**   Tag memory is divided into four banks. A bank may have read-only, write once, and read/write parts within it. This provides better tag security and application flexibility.

  Tag memory has four banks numbered from 0 to 3. Bank 0, called Reserved Memory, holds Kill and Access passwords, each 32 bits long; Bank 1, called EPC Memory, holds the EPC including the EPC header or Asset Family Identifier (AFI) based on ISO 15961, Protocol Control (PC) and Cyclic Redundancy Check (CRC-16); Bank 2, called Tag Identification Memory, holds a tag identifier based on ISO 15963, this bank is not rewritable; and Bank 3, called User Memory, is optional and holds user defined data.
- **Q Algorithm**   Provides faster resolution of tag collision and increased security for communication between tag and reader.
- **More robust tag communication design**   Reduces potential for ghost reads and entry of erroneous data into the application.

Local Lingo

Dense reader environment   *Dense reader environment* and *dense interrogator environment* have equal meanings.

EPC Tag Data Standards, Version 1.3

This standard defines EPC tag data formats for Generation 2 tags. It defines how the EPC is encoded on the tag and how it is encoded for use in the information systems layers of the EPC Systems Network. The standard includes specific encoding schemes for EPC General Identifier (GID). It also defines encoding of six other numbering systems used in global trade:

- **SGTIN (Serialized GTIN)**   Serialized EAN.UCC Global Trade Item Number
- **SSCC**   EAN.UCC Serial Shipping Container Code
- **GLN**   EAN.UCC Global Location Number
- **GRAI**   EAN.UCC Global Returnable Asset Identifier
- **GIAI**   EAN.UCC Global Individual Asset Identifier
- **DoD**   US Department of Defense Number

The standard defines two versions for each of these six numbering schemes. One version has a length of 64 bits, while the other has 96 bits. Table 10-1 shows the GID-96 numbering scheme

Travel Advisory

The 64-bit EPC is currently approaching its sunset in favor of 96-bit EPC.

The EPC tag encodings include a header field followed by one or more value fields. The header field defines the overall length and format of the value fields. The value fields contain a unique EPC identifier and optional filter value. The EPC uniform resource identifier (URI) encodings provide the means for applications software to process EPC. The encoding schemes are independent of the tag types used. This standard defines four different categories of URI:

- **URIs for pure identities, sometimes called canonical forms**   These contain only the unique information that identifies a specific physical object and are independent of tag encodings.
- **URIs that represent specific tag encodings**   These are used in software applications where the encoding scheme is relevant, as when commanding software to write a tag.

TABLE 10.1	Format of EPC GID-96			
	Header	General Manager Number	Object Class Number	Serial Number
GID-96	8 bits	28 bits	24 bits	36 bits
Decimal capacity	0011 0101 (actual value)	268,435,455	16,777,215	68,719,476,736 ~68 billion

- URIs that represent patterns or sets of EPCs   These are used when instructing software how to filter tag data.

- URIs that represent raw tag information   These are generally used only for error-reporting purposes.

Travel Advisory

The older standard defining the tag data formats for Generation 1 tags is the EPC Generation 1 Tag Data Standards Version 1.1 Rev.1.27 (May 2005).

Exam Tip

You will not be asked about the URIs, but they are good to know in your RFID endeavors.

This concludes the discussion of important RFID standards. Many more standards are used in design, manufacturing, deployment, and maintenance of RFID systems. More detailed information regarding all the standards is available from the Web sites of the ISO and EPCglobal or from published documents.

Objective 10.2 Regulations

RF signals (waves) travel almost forever and pass through many solid materials (though their strength does get reduced due to path loss and absorption). Therefore, they cannot be easily contained within a desired space, nor can we ignore the effects of radio signals transmitted by devices located in a long distance, even thousands of feet, away. These devices may be inside other buildings and may not be visible, but they can interfere with your RF system, and your system can interfere with them. These types of interferences affect the performance of an RF system. They may, depending on their relative strength, reduce the read range of your system or render the system inoperable.

Since RF signals do not obey the normal property boundaries created by humans, they create an environment like the Wild West of older, where bullets shot by good or bad guys fly forever, passing through some walls and ricocheting off others. Anybody in the path of these bullets can be injured. To prevent this type of chaos with RF signals, a system of "laws and sheriffs" is needed to make sure that nobody gets hurt. In the RFID arena, various governmental agencies called *regulatory authorities* provide this system.

These regulatory authorities define who can use what RF for what purposes and how. Consider, for example, radio and TV stations. Each station, in addition to other attributes, is assigned a geographic location, radio frequency band, and strength of the radio signal they can emit from their antennas. This prevents one station from interfering with the other stations, and as a result everybody is able to listen to their station of choice. These regulations are laws of the land and must be obeyed. Unlike standards, which are optional, regulations are mandatory, and if they are not followed properly, legal consequences will ensue. If you do not follow standards, your devices may not interoperate with other devices or some customers may not buy your products, but you can still continue doing business, albeit with financial consequences. If you do not follow the regulations, government authorities may fine you, and the repeated violations may result in more severe penalties such as not being allowed to make or use RF devices—or even a jail term.

Another reason to control RF transmission is to avoid injuries to humans and animals. RF devices transmit and receive RF energy. For example, Wi-Fi devices operating at 2.4 GHz range, which is the same frequency a microwave oven uses, can seriously damage human tissue if they transmit a signal at a very high level. In a low strength, they are considered harmless. A regulation must be established as to the level of safe exposure, and some mechanism has to be created to certify and monitor compliance.

RF regulations typically vary from one country to another or from one region to another, due to the legacy usage of various portions of the RF spectrum. In the technology's early years, various countries or regions assigned different chunks of the RF spectrum for different uses. No worldwide standard was available, and even now none exist for many parts of the spectrum. Finding a worldwide RF range for new RF applications is a problem. For example, in the US, the UHF RFID systems are allocated a frequency range of from 902 to 928 MHz, but in European countries, that range was already assigned to other uses and is therefore not available. In Europe, UHF RFID systems are assigned a frequency range of 865 to 868 MHz. As a result, a tag designed for the US will have problems being read in Europe and vice versa. To overcome these problems, RFID systems must be designed to incorporate all the frequency ranges (within the UHF band) used all over the world. This has been accomplished by EPCglobal Gen 2 and ISO 18000-6C standards. Readers and tags designed according to these standards will interoperate anywhere in the world.

For UHF RFID tags and readers, regulations encompass the following major factors:

- **RF field power**   Effective isotropic radiated power (EIRP) in watts.
- **Bandwidth usage**   The frequency range allocated.

- **Channels and channel spacing** How the allocated frequency range is divided into channels to incorporate reader to reader interference.
- **Duty cycle** The percentage of time a reader can actively transmit.

Some of the main bodies in various countries governing frequency allocations and regulations for RFID are presented in Table 10-2.

Regulatory Organizations

This section discusses some of the leading RF regulatory bodies that control use of the radio spectrum in many regions of the world. These organizations are created by one country or a regional consortium of many countries.

International Regulations: ITU

The International Telecommunication Union was established in Paris as the International Telegraph Union in 1865 and is today the world's oldest international organization. It became a specialized agency of the United Nations in 1947. Its main tasks include standardization, allocation of the radio spectrum, and organizing interconnection arrangements among countries to allow international phone calls—in which regard it performs for telecommunications a similar function to what the UPU performs for postal services. It is one of the specialized agencies of the United Nations, and has its headquarters in Geneva, next to the main United Nations campus. It provides a forum in which its 189 member states, some 650 sector members, and more than 90 associates can cooperate for the improvement and rational use of telecommunication worldwide.

TABLE 10.2 Main Regulatory Organizations Related to RFID

Worldwide	International Telecommunication Union (ITU)
USA	Federal Communications Commission (FCC)
Canada	Department of Communication (DOC)
Europe	European Radiocommunications Office (ERO) and European Telecommunications Standards Institute (ETSI)
Japan	Ministry of Public Management, Home Affairs, Post and Telecommunications (MPHPT) or Ministry of Internal Affairs and Communications (MIC)
China	Standardized Administration of China (SAC)
Singapore	Infocomm Development Authority (IDA) of Singapore
Oceania	Australian Communication Authority, New Zealand Ministry of Economic Development

The ITU is made up of three bureaus:

- Telecommunications Bureau (ITU-T)
- Radiocommunications Bureau (ITU-R)
- Development Bureau (ITU-D)

The international standards that are produced by the ITU are referred to as "Recommendations" (with the word ordinarily capitalized to distinguish its meaning from the ordinary sense of the word *recommendation*). Due to the ITU's longevity as an international organization and its status as a specialized agency of the United Nations, standards promulgated by the ITU carry a higher degree of formal international recognition than those of most other organizations that publish technical specifications of a similar form. ITU has divided the world in three different regions, as shown in Figure 10-2.

Regulations in Europe: ETSI

The European Telecommunications Standards Institute is an independent, not-for-profit organization located in Sophia Antipolis, France. Its mission is to produce telecommunications standards, and it is officially responsible for standardization of Information and Communication Technologies (ICT) within Europe. It develops a wide range of standards and provides interoperability testing services. ETSI has 654 members from 59 countries inside and outside Europe, including manufacturers, network operators, administrations, service providers, research bodies, and users. ETSI's prime objective is to support global harmonization by providing a forum in which all key players can contribute actively. ETSI is officially recognized by the European Commission and the EFTA (European Free Trade Association) secretariat. Two of the most important RFID regulations created by ETSI are EN 300-220 and EN 302-208.

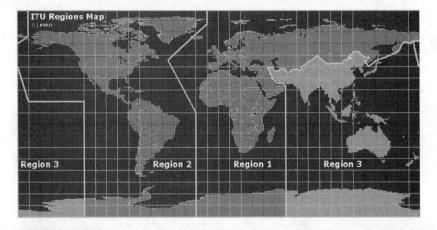

FIGURE 10.2   ITU regions

ERO

The European Radiocommunications Office is the permanent office that supports the Electronic Communications Committee (ECC), formerly the European Radiocommunications Committee (ERC). The ECC is the telecommunications regulation committee for the European Conference of Postal and Telecommunications Administrations (CEPT), representing 46 countries and covering most of the geography of Europe. The main task of the ECC is to develop radiocommunications policies. It coordinates frequency regulation and technical matters for the allocation and utilization of the 9 kHz to 275 GHz frequency range.

The ERO publishes and distributes ECC Decisions and Recommendations. Recommendations related to the use of short range devices (SRDs) are found in ERC Recommendation 70-03. The ERC Decision ERC/DEC(01)04 addresses nonspecific SRDs operating in 868.0–868.6 MHz, 868.7–869.2 MHz, 869.40–869.65 MHz, and 869.7–870.0 MHz.

Regulations in the United States: FCC

The Federal Communications Commission is an independent agency of the US government, directly responsible to Congress. It was established by the Communications Act of 1934 and is charged with regulating interstate and international communications by radio, television, wire, satellite, and cable. The FCC's jurisdiction covers the 50 states, the District of Columbia, and US possessions. The FCC is directed by five commissioners appointed by the president and confirmed by the Senate. The president designates one of the commissioners to serve as chairperson. The commissioners supervise all FCC activities, delegating responsibilities to staff units and bureaus. The commission includes six operating bureaus and ten staff offices within the FCC.

One of the bureaus is the Wireless Telecommunications Bureau (WTB), which handles nearly all FCC domestic wireless telecommunications programs, policies, and outreach initiatives. Wireless communications services include amateur, cellular, paging, broadband PCS, public safety, and more. FCC regulates RF usage via regulations that are laws in the United States, and all RF usage within the US must confirm to these regulations.

In the US, FCC Part 15 regulates the use of RF devices. Its subpart C deals with intentional (fixed) radiators, and Section 15.247 specifically deals with RFID devices operating at UHF frequencies within the bands of 902–928 MHz, 2400.0–2483.5 MHz, and 5725–5850 MHz. UHF can be used unlicensed for 908–928 MHz, but restrictions exist for transmission power and channel usage.

RFID Regulations

This section discusses some of the important regulations created by the organizations and agencies discussed in the previous section. These regulations define in detail how RF can be used and for what purposes.

FCC Part 15

RFID devices operating at UHF frequencies are allowed for operation in the Industrial, Scientific, and Medical (ISM) bands under conditions defined in FCC Part 15 rules, Section 15.247. Section 15.247 defines operation within the bands 902–928 MHz, 2400.0–2483.5 MHz, and 5725–5850 MHz. The 902–928 MHz band offers optimum range of operation and is usually preferred for supply chain applications. Part 15–compliant RFID systems typically use a frequency-hopping spread spectrum modulation technique to benefit from maximum reader transmitted power allowances. Part 15–compliant UHF readers can operate at a maximum transmitted power of 1 watt, or up to 4 watts with a directional antenna, if they hop across a minimum of 50 channels. For more details, you can download Part 15 rules from the FCC Web site.

FCC Section 15.247 has nine different paragraphs that provide specifications for using RF devices, as shown in Table 10-3.

FCC Section 15.247(b) deals with regulations for frequency hopping within 902–928 MHz:

- If the 20 dB bandwidth of the hopping channel is less than 250 kHz, the system shall use at least 50 hopping frequencies and the average time of occupancy on any frequency shall not be greater than 0.4 seconds within a 20-second period.

- If the 20 dB bandwidth of the hopping channel is 250 kHz or greater, the system shall use at least 25 hopping frequencies and the average time of occupancy on any frequency shall not be greater than 0.4 seconds within a 10-second period.

- The maximum allowed 20 dB bandwidth of the hopping channel is 500 kHz.

TABLE 10.3	Description of FCC 15.247 Paragraphs
Paragraph	**Specifications Relate to**
a	Compliances for frequency hopping for conventional and digitally modulated intentional radiators
b	Maximum peak conducted output power of intentional radiators
c	Operations with directional antenna gains greater than 6 dBi
d	Limits on RF power emitted outside the frequency band
e	Power spectral density of digitally modulated systems
f	Hybrid systems
g	Frequency hopping spread spectrum and channel hopping
h	Incorporation of intelligence within a frequency hopping spread spectrum
i	Limits of exposure to radio frequency energy levels

FCC Section 15.247(b) deals with maximum peak conducted output power of intentional radiators:

- The conducted output power limit for various frequency bands is based on the use of antennas with directional gains that do not exceed 6 dBi.
- If transmitting antennas of directional gain greater than 6 dBi are used (see Section 15.247(c)), the conducted output power from the intentional radiator shall be reduced below the stated values, as appropriate, by the amount in dB that the directional gain of the antenna exceeds 6 dBi.
- For systems using digital modulation in the 902–928 MHz, 2400.0–2483.5 MHz, and 5725–5850 MHz bands, the maximum peak conducted output power of intentional radiators is 1 watt.
- As an alternative to a peak power measurement, compliance with the 1 watt limit can be based on a measurement of the maximum conducted output power.

ETSI EN 300-220

This standard applies to SRD radio transmitters and receivers. It covers transmitters in the range from 25 to 1000 MHz and with power levels ranging up to 500 mW, and receivers in the range from 25 to 1000 MHz. It defines the technical characteristics for radio equipment. It defines product family information that may be completely or partially superseded by specific standards covering specific applications. It applies to a RF output connection and/or an integral antenna and to alarms, identification, telecommands, and telemetry applications.

When selecting parameters for new SRDs, which may have inherent safety of human life implications, providers and users should pay particular attention to the potential for interference from other systems operating in the same or adjacent bands. It covers fixed stations, mobile stations, and portable stations, and requirements are provided for the different frequency bands, channel separations, modulations, and other parameters. Some of the parameters defined are frequency range 869.4 to 869.65 MHz, bandwidth 0.25 MHz, maximum allowable power 0.5 watts ERP, one channel, and 10 percent duty cycle. National restrictions may apply to the regulation.

ETSI EN 302-208

The key features of ETSI regulation EN 302-208 are as follows:

- Shared operation in band 865–868 MHz at transmit powers up to 2 watts effective radiated power (ERP)
- Mandatory "listen before talk" function
- Operation in sub-bands of 200 kHz

- Power levels of 100 mW, 500 mW, and 2 W ERP
- Mandatory listen time of more than 5 ms before each transmission
- Maximum period of continuous transmission of 4 seconds
- Pause of 100 ms between repeated transmissions on the same sub-band, or the interrogator will move immediately to another vacant sub-band

This regulation allows European RFID interrogators operating in the UHF band to perform nearly as well as UHF interrogators operating under FCC rules in the United States. It provides an additional frequency range, compared to EN 302-200, from 865 to 868 MHz for RFID interrogator operation. This increases the spectrum band from 250 kHz to 3 MHz. The number of channels on which readers can broadcast has been increased from 1 to 15. The new band is divided into three sub-bands (see Figure 10-3). Under the old regulations EN 300-220, UHF readers were restricted to half a watt of ERP. The new regulations allow them to emit up to 0.1 watt ERP between 865 and 865.5 MHz, 2 watts ERP between 865.6 and 867.6 MHz, and 0.5 watt ERP between 867.6 and 868 MHz.

The duty cycle restrictions are replaced with the listen before talk (LBT) algorithm. The interrogator can stay on a selected channel for up to 4 seconds; it must then stop emitting energy for at least 0.1 second to provide other devices with the opportunity to use the channel. The interrogator could also switch immediately to any other unoccupied channel and transmit. The interrogators without LBT capabilities are limited to a 0.1 percent duty cycle. The data rate of this regulation is less than in the United States. This is because only 3 MHz of the spectrum is available in Europe for RFID, while 26 MHz is available in the United States.

Regulations in Canada

In the 902–908 MHz UHF band, Canadian standards are similar to US regulations. The Canadian Cattle Identification Agency (CCIA) or recently named Canadian Livestock Identification Agency (CLIA) has recommended that all animal tags be replaced with RFID tags. This has been backed by the Canadian cattle industry so it can continue to meet domestic and international requirements for

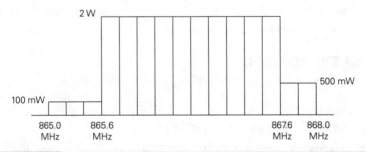

FIGURE 10.3   Band division and allowed transmitted power (ERP) according to ETSI EN 302-208

animal health and food safety through an efficient trace-back and age verification process using RFID technology. Removing a CCIA tag is a serious offense in Canada. CCIA requires that all approved RFID tags are yellow in color.

Regulations in Japan

The Ministry of Public Management, Home Affairs, Post and Telecommunications (MPHPT), whose English name has been recently changed to Ministry of Internal Affairs and Communications (MIC), is responsible for formulating policies related to communications, which includes setting up of RFID standards within Japan. MIC has agreed to open up the UHF spectrum range of 952–954 MHz for RFID use. High-powered passive tag systems can use an antenna power between 10 mW to 1 W and an antenna gain of 6 dBi, which gives maximum power equivalent to 4 W EIRP. Users must obtain a license to use this RFID system. For low-powered systems of up to 10 mW, no user license is required. You may also see a band of 950–956 MHz specified for Japanese regulation, since Japanese regulations are not yet 100 percent established.

Regulations in China

In the last few years, China has been in the process of developing its own RFID standards to be in line with global standards. Standardization in China starts with the Standardization Administration of China (SAC), a ministry-level organization in the Chinese government. Under this organization are many Technical Committees focused on specific technology and business areas. The RFID National Standards Working Group, under China National Registry of Product and Service Codes (NPC), is responsible for standardizing hardware-related issues including frequency allocation, bandwidth, and tag-to-reader communication. The Article Numbering Center of China (ANCC) is focused on the tag encoding format making sure that goods tagged in China are compliant with the Electronic Product Code (EPC) coding scheme.

Regulations in India

The Indian government arm that regulates applications of radio waves, the Department of Telecommunication (under the Ministry of Communications and Information Technology), has recently delicensed the spectrum in the 865–867 MHz band for use by RFID devices. The regulation on the use of wireless equipment in the band 865–867 MHz specifies that no license is required by any person to establish, maintain, work, possess, or deal in RFID on noninterference, nonprotection, and nonexclusive bases, in the frequency band 865–867 MHz with maximum 1 W transmitter power, 4 W ERP, and 200 kHz carrier bandwidth.

Regulations in Australia

The UHF band for RFID operations in Australia is 918–926 MHz with a power of 1 W EIRP. The cattle, food, and beverages industry is supporting Australia's drive to integrate internationally with the RFID technology. The Australian Communication Authority, responsible for setting RFID standards within Australia, has been considering the globally accepted industry-driven standards for the EPC.

Regulations in Singapore

The Infocomm Development Authority (IDA) of Singapore regulates radio frequencies allocated for various applications within Singapore. The IDA has recently announced the new spectrum allocation and power limits for RFID usage in Singapore in the UHF band.

Singapore allows use of the 866–869 MHz and 923–925 MHz bands within the UHF spectrum for RFID activities. The power limit for both bands is 0.5 W. It is expected that the power limit for the 923–925 MHz band will be increased to 2 W for RFID devices. Table 10-4 summarizes key regulatory parameters in various parts of the world.

Objective 10.3   Industry Mandates

Mandates are not part of the RFID+ certification exam but since there is so much talk about them in the RFID industry, this section provides some information about them. Mandates are created by large organizations that buy goods from many suppliers located around the world. Their goal is to create, by using RFID, a more efficient supply chain and thereby reduce costs. As with standards, compliance with mandates is not required, but noncompliance may affect your relationship with the mandating organization—for example, you may not be able to do business with them. All the mandates require use of UHF systems, and almost all of them now require the use of Gen 2 (ISO 18000-6C) tags.

TABLE 10.4	Global Regulatory Requirements for UHF Band				
Country	**Regulators/ Regulations**	**Frequency**	**Bandwidth**	**Channel Spacing**	**Maximum Power**
USA	FCC Part 15, Section 247	902–928 MHz	26 MHz	52 channels of 500 kHz spacing	4 W EIRP
Europe	CEPT/ETSI 302-208	865.0–868.0 MHz	3 MHz	15 channels of 200 kHz spacing	0.1 W ERP 2.0 W ERP 0.5 W ERP
Japan	MIC	952–954 MHz	2 MHz		4 W EIRP
India	DOT	865–867 MHz	2 MHz	10 channels of 200 kHz spacing	4 W EIRP
Singapore	IDA	923–925 MHz	2 MHz		2 W EIRP

All the mandates are phased in over a time period of more than one year and include using RFID tags on pallets and cases of items shipped. No item-level tagging has been mandated yet. Mandates also require some sort of electronic document to be sent to the customer with a list of all the tag numbers shipped. Many suppliers, particularly small companies, have started placing RFID tags on cases and pallets just before shipping them. This is typically called "tag and ship" or "slap and ship." The companies using slap and ship are incurring the cost of the RFID system but are not getting any benefits from it. One advantage of mandates is that they have accelerated the use of RFID. Without the mandates, many companies would not have quickly adopted RFID.

CHECKPOINT

✔**Objective 10.1: Standards** To design and implement a well-functioning RFID system, you must comply with related standards that can be international, regional, and national, as well as regulations for the particular country and operation. Standards define air interface protocols, data protocols, and capabilities and functions of various types of RFID and RF systems.

The air interface protocol defines the rules of communication between an interrogator and a tag. Tag data format or the tag data protocol defines the structure and encoding of data to a tag.

Main international standards that need to be observed for RFID are ISO 14443 for proximity cards (HF), ISO 15693 for vicinity cards (HF), ISO 15963 for ID of RF tags, and ISO/IEC 18000 air interface protocol for RFID systems (LF, HF, UHF active and passive, microwave). Other important standards are national standards created by organizations such as ANSI in the United States and BSI in the UK. Compliance with standards is important to ensure proper function, interoperability, and safety. Regulations are developed by each country or region (such as EU). In the US, the regulations for RFID are developed by the FCC (remember part 15.247); in the EU, it is ETSI (remember EN 300 220 and EN 302 208). Other national regulatory organizations are SAC in China, MIC in Japan, DOT in India, IDA in Singapore, to name a few. Regulation compliance is mandatory and you can be fined for noncompliance.

✔**Objective 10.2: Regulations** Regulations deal mainly with transmitted power limits to prevent interference and ensure human safety. In the US, RFID is regulated by FCC, where FCC Part 15 allows up to 4 W EIRP with gained antenna and hopping across 50 channels in a frequency band 902–928 MHz. In Europe, RFID is regulated by ETSI's two main standards. ETSI EN 300 220 regulates frequency band 869.4–869.65 MHz and allows 0.5 W ERP with 10 percent duty cycle. Newer EN 302 208 adds the band of 865–858 MHz, and allows ERP up to 2 W and LBT. Japan allows current RFID operations at 952–954 MHz with 4 W EIRP.

The transmitted power of RFID and other radiating systems is limited to ensure human safety when exposed to these signals. The limits are set by various regulatory organizations (such as FCC and ETSI). If your system operates under local regulations and complies with standards, it is safe for installation, handling, and operation by human personnel as far as exposure to RF waves.

✔**Objective 10.3: Industry Mandates**   Mandates are created by large organizations that buy goods from many suppliers located around the world. Their goal is to create, by using RFID, a more efficient supply chain and thereby reduce the cost. Noncompliance with a mandate may affect your relationship with the mandating organization.

REVIEW QUESTIONS

1. Which international standard describes the air interface protocol for passive UHF RFID systems?

    A. ISO/IEC 18000-6

    B. ISO/IEC 14443

    C. EPCglobal Standard version 1.27

    D. ETSI EN 302-208

2. Which organization globally assigns frequencies of operation to various regions and technologies?

    A. ISO

    B. ETSI

    C. ITU

    D. FCC

3. To which ITU region does the United States belong?

    A. Region 1

    B. Region 2

    C. Region 3

    D. Region 4

4. According to ETSI EN 302-208 regulation, what is the maximum allowable amount of power (ERP) that can be transmitted from a UHF reader?

 A. 1 watt

 B. 2 watts

 C. 0.25 watt

 D. 4 watts

5. What is the current frequency band used for UHF RFID operations in Japan?

 A. 902–928 MHz

 B. 866–868 MHz

 C. 852–855 MHz

 D. 952–954 MHz

6. What is the function of a header in the EPC number?

 A. Header specifies the size of the EPC memory.

 B. Header specifies the air interface protocol of the EPC tag.

 C. Header specifies the type, format, and length of the EPC number.

 D. Header specifies the unique instance of the tag.

7. Which ISO/IEC standard specifies a unique ID of RF tags within ISO 18000-6C?

 A. 15693

 B. 15961

 C. 15963

 D. 18000-6

8. Which standard describes the air interface protocol for Generation 2 tags? (Select two answers.)

 A. EPC standard version 1.27

 B. EPC standard version 1.0.9

 C. ISO/IEC 18000 Part 6C

 D. ISO/IEC 18000 Part 3B

9. What is the rule for operation for FCC Part 15–compliant UHF readers?

 A. UHF readers can operate at a maximum transmitted power of 1 watt, or up to 4 watts with a directional antenna, if they hop across a minimum of 20 channels.

B. UHF readers can operate at a maximum transmitted power of 0.5 watt, or up to 4 watts with a directional antenna, if they hop across a minimum of 25 channels.

C. UHF readers can operate at a maximum transmitted power of 1 watt, or up to 4 watts with a directional antenna, if they hop across a minimum of 50 channels.

D. UHF readers can operate at a maximum transmitted power of 1 watt, or up to 4 watts with an isotropic antenna, if they hop across a minimum of 50 channels.

10. Which region of the world requires duty cycling under certain conditions for operation of UHF RFID readers?

A. The United States

B. Europe

C. India

D. Singapore

REVIEW ANSWERS

1. **A**   ISO/IEC 18000-6 standard describes the air interface protocol for passive UHF RFID systems.

2. **C**   ITU globally assigns frequencies of operation to various regions and technologies.

3. **B**   United States belongs to Region 2.

4. **B**   The maximum allowed transmit power (ERP) in Europe according to regulation ETSI EN 302-208 is 2 watts.

5. **D**   The current frequency band used for the UHF RFID operations in Japan is 952–954 MHz.

6. **C**   Header specifies the type, format, and length of the EPC number.

7. **C**   ISO/IEC 15963 IEC standard specifies a unique ID of RF tags within ISO 18000-6C.

8. **B C**   EPC standard version 1.0.9 and ISO/IEC 18000 Part 6C standards describe the air interface protocol for Generation 2 tags.

9. **C**   UHF readers can operate at a maximum transmitted power of 1 watt, or up to 4 watts with a directional antenna, if they hop across a minimum of 50 channels.

10. **B**   Europe requires duty cycling under certain conditions for operation of UHF RFID readers.

About the CD-ROM

Mike Meyers' Certification Passport CD-ROM Instructions

The CD-ROM included with this book comes complete with MasterExam and the electronic version of the book. The software is easy to install on any Windows 98/2000/XP/Vista computer and must be installed to access the MasterExam feature. You may, however, browse the electronic book directly from the CD without installation. To register for a second bonus MasterExam, simply click the Online Training link on the Main Page and follow the directions to the free online registration.

System Requirements

Software requires Windows 98 or higher and Internet Explorer 5.0 or above and 20MB of hard disk space for full installation. The electronic book requires Adobe Acrobat Reader.

Installing and Running MasterExam

If your computer CD-ROM drive is configured to auto run, the CD-ROM will automatically start up upon inserting the disk. From the opening screen you may install MasterExam by pressing the *MasterExam* buttons. This will begin the installation process and create a program group named "LearnKey." To run MasterExam use START | PROGRAMS | LEARNKEY. If the auto run feature did not launch your CD, browse to the CD and click the "LaunchTraining.exe" icon.

MasterExam

MasterExam provides you with a simulation of the actual exam. The number of questions, the type of questions, and the time allowed are intended to be an accurate representation of the exam environment. You have the option to take an open book exam, including hints, references, and answers, a closed book exam, or the timed MasterExam simulation. When you launch MasterExam, a digital clock display will appear in the upper left-hand corner of your screen. The clock will continue to count down to zero unless you choose to end the exam before the time expires.

Electronic Book

The entire contents of the Study Guide are provided in PDF. Adobe's Acrobat Reader has been included on the CD.

Help

A help file is provided through the help button on the main page in the lower left-hand corner. Individual help features are also available through MasterExam.

Removing Installation(s)

MasterExam is installed to your hard drive. For BEST results for removal of programs use the START | PROGRAMS | LEARNKEY | UNINSTALL options to remove MasterExam.

LearnKey Technical Support

For technical problems with the software (installation, operation, removing installations), and for questions regarding online registration, please visit www.learnkey.com or e-mail techsupport@learnkey.com.

Content Support

For questions regarding the technical content of the electronic book or MasterExam, please visit www.osborne.com or e-mail customer.service@mcgraw-hill.com. For customers outside the 50 United States, e-mail: international_cs@mcgraw-hill.com.

Career Flight Path

APPENDIX B

In my former life as a LAN/WAN professional, I used to ask the question, Why should I certify? Having learned my lessons on the "street," I would watch the "pedigree" engineers struggle with a problem, only to give up while the "street" engineer would step in and solve the problem but never receive credit for it. Years later, after I was over my embitterment about this, I saw (with help from my super-certified brother) the answer. Plain and simple: it's about credibility.

As a real-world engineer you can solve the problem, but will you ever get the chance? Taking the time to test out of a certification program tells your employer that you not only know what you are talking about, but you are secure enough in your knowledge to prove it.

As the RFID industry gains momentum, more and more consulting opportunities present themselves. Individuals pursuing these opportunities find immense value in certification of knowledge, as certification often is the very reason they are selected for the job. Many of my students are already well versed in consulting and project management and use the RFID+ certification as another tool in their vast array of already established solutions capabilities.

RFID+ certification was created to address the needs of this emerging market. Once certified, a RFID+ certified technician should have the core knowledge to install and troubleshoot RFID reader installations, participate in RFID solution design, identify various RFID tag and reader technologies visibly, decide what technology best addresses a specific scenario, and contribute as a value-add to any RFID solutions team. Many RFID+ certified technicians are former LAN/WAN professionals or supply-chain support professionals who are looking to utilize the skills they have as they acquire more.

As this certification has been designed as a vendor-neutral, entry-level exam focused on RFID systems implementation, you may want to consider continuing your education in areas that are closely related to RFID systems or where the RFID systems plug in. This means that knowledge of networking, computer hardware and programming, as well as vendor-specific RFID hardware and software could lead you on the interesting career path of an RFID professional.

To achieve some credibility in those areas, consider following certifications:

* CompTIA Network+ Certification
* CompTIA A+ Certification
* CWNP–Certified Wireless Network Administrator (CWNA)
* CompTIA Linux+ Certification
* Vendor-specific certifications

CompTIA Network+ Certification   CompTIA Network+ Certification is one of the recommended certifications you should master before you take the RFID+ exam. If you haven't been certified in Network+, I suggest you do it before taking the RFID+ exam. This certification will demonstrate your skills in installation, configuration, and troubleshooting of basic networking hardware, protocols, and services, which is essential knowledge for design and deployment of RFID networks.

CompTIA A+ Certification   CompTIA A+ Certification shows that you have the ability to deal with issues that can come up during installation, configuration, diagnosis, preventive maintenance, and basic networking of computer systems, which are used as host systems for the RFID installations. By this increased competency, you will not have to depend on local IT folks when it comes to smooth cooperation of an RFID system with an internal IT system.

CWNP–Certified Wireless Network Administrator (CWNA)   CWNA provides an in-depth testing of your knowledge of wireless communications, radio frequency, and local area networking, which are essential for deployment of 802.11 wireless communication devices. This is especially important since this type of data transfer has become more and more popular with mobile as well as fixed RFID applications.

CompTIA Linux+ Certification   Since many back-end systems as well as RFID readers are running on a Linux platform, you should acquire the knowledge and certification of Linux, which demonstrates your ability of fundamental management of Linux systems from the command line, knowledge of user administration, understanding of file permissions and software configurations, and management of Linux-based clients, server systems, and security.

Vendor-Specific Certifications   In your career as an RFID professional, you will encounter devices made by various manufacturers. They will all work on the same principle, however, though their features and capabilities may be slightly different. To sell, use, and troubleshoot the equipment properly, you should receive specific certifications from vendors such as Symbol, Printronix, ThingMagic, Zebra, Sun Microsystems, Intermec, and others.

Glossary

active tag   Active tags have an internal battery to power them and an active transmitter to transmit data. Some active tags contain replaceable batteries for years of use; others are sealed units. Active tags generally can transmit data over a longer distance, are considerably larger than passive tags, and have a limited operational lifespan. They are generally used for tracking expensive items over long ranges.

agile reader   An RFID reader that can read tags operating at different frequencies and/or using different protocols to communicate between the tags and readers.

air interface protocol   The standards that define how RFID tags and readers communicate using radio waves.

alignment   The orientation of the tag to the reader in pitch, rolls, and yaws. (*See* orientation.)

American National Standards Institute (ANSI)   American technical standards body that also represents United States to the International Organization for Standardization.

amplitude   The maximum absolute value of a periodic, trigonometric sine curve measured along its vertical axis.

amplitude modulation   A method of transmitting information by varying (modulating) the amplitude of the carrier wave.

amplitude shift keying   A method of transmitting information using various levels of amplitudes of a sinusoidal wave. For a digital system, two amplitude levels are usually used: a level to represent a digital one and a lower level to represent digital zero. This is a special case of amplitude modulation.

antenna   A transducer made of a conductive material that converts alternating current (AC) supplied through an antenna's feed line to radio waves and also converts received radio waves to AC. Radio waves radiate outward from the antenna. RFID systems use two types of antennas: one connected to the reader and another on the tag. RFID systems use two types of antennas: one connected to the reader and another on the tag.

antenna gain   The ratio of the amount of energy radiated in one direction and the energy radiated by a reference antenna in the same direction when driven by the identical input. Gain is generally specified in decibels (dB). Usually we are interested in the maximum gain—the direction in which the antenna is radiating the most power.

anti-collision   In the context of RFID, refers to different ways of keeping radio waves from one tag from interfering with radio waves from another tag. Anti-collision algorithm is used to collect data from multiple RFID tags present at the same time in a reader's interrogation zone without interference.

Association for Automatic Identification and Mobility   A trade association of companies that globally provide products and services related to automatic identification, data collection, networking, and information management systems.

ATR (Answer To Reset)   A data string returned by a smart card when it is powered on.

attenuation   Reduction of RF energy from an RFID tag or reader; water absorbs UHF energy, causing signal attenuation.

attenuator   A device that attaches to a transmission line and reduces the power of the RF signal as the signal travels through the cable from the reader to the antenna. Attenuators usually work by dissipating the RF energy into heat.

Auto-ID Center   A non-profit collaboration between private companies and academia that pioneered the development of an Internet-like infrastructure to track goods globally through the use of RFID tags carrying Electronic Product

Codes. The center closed its doors in September 2003. The organization EPCglobal now continues this work.

Auto-ID Labs   Nonprofit research labs, headquartered at the Massachusetts Institute of Technology, that continue to perform primary research into the development of Electronic Product Code (EPC) and related technologies.

Automatic Identification (Auto-ID)   A broad term used for the methods of collecting data and entering it into computer systems without human involvement. Technologies normally considered part of Auto-ID include barcodes, biometrics, RFID, and voice recognition.

Automatic Identification and Data Capture (AIDC)   A broad term that covers methods of identifying objects, capturing information about them, and entering it directly into computer systems without human involvement.

auto-talking   A transponder is auto-talking when it starts sending its memory content as soon as it enters an RF field from a read/write unit, without the need of a special command.

back channel   Communication method used by the signal returning from the RFID tag back to the interrogator. Also known as reverse channel or reverse link.

backscatter   A method of communication by which a tag reflects the carrier wave received from the interrogator. To transfer information, the tag modulates the amplitude of the reflected wave. Passive and semi-passive tags use this method to send information to the interrogator.

barcode   A way of encoding information using black-and-white stripes or patches. Stripes are used by linear barcodes and patches are used by two-dimensional barcodes. The most common barcode is Universal Product Code (UPC) used on most grocery and retail items. Barcodes are read using laser scanners and require a line-of-sight.

base station   An electronic device made up of an RF transmitter, a receiver, and an antenna. It is used to communicate with RFID transponders and contactless smart cards. It connects systems such as access control terminals, ticket vending machines, and car immobilizer units to host computers.

battery-assisted tag (BAT)   An RFID tag that has a battery but does not have an active transmitter. It communicates using the same backscatter technique used by passive tags. It uses the battery to run the circuitry on the integrated circuit and sometimes an on-board sensor. BATs have a longer read range than passive tags. They are sometimes called semi-passive RFID tags or battery-assisted passive (BAP) tags.

baud In a digitally modulated signal, baud measures symbol rate: that is, the number of distinct symbol changes (signaling events) made to the transmission medium per second. The baud rate is not the same as bit rate because one symbol may carry more than 1 bit of data. Sometimes many symbols are used to convey 1 bit of data.

beacon An active RFID tag that is programmed to wake up and broadcast its signal at a set interval.

bistatic reader An interrogator that uses two antennas to interrogate the tags: one to transmit the signal and another to receive. The two antennas, transmit (Tx) and receive (Rx), form one interrogation port. An interrogator may have one or more such ports.

BNC connector A male type of connector used with coaxial cables. It has a center pin connected to the center conductor and a metal tube connected to the outer cable shield. An outer rotating ring is used to lock the cable to a female connector. It was named after its bayonet mount locking mechanism and its two inventors Neill and Concelman. Over the years it has picked up other names such as bayonet navy connector, baby N connector, barrel nut connector, bayonet nipple connector, British naval connector, and British national connector.

capacitor An electric circuit element used to store a charge temporarily. A capacitor usually consists of two metallic plates separated and insulated from each other by a dielectric substance.

card operating system The software program stored in the smart card IC that manages the basic functions of the card, such as communication with the terminal, security management, and data management in the smart card file system.

carrier frequency The base or central frequency of a sinusoidal wave used to carry (transmit) information using radio waves, electric current through wires, or light waves through optical fibers. This frequency is modulated by varying its amplitude, frequency, or phase to carry information.

checksum A code added to the contents of a block of data stored on an RFID integrated circuit that can be checked before and after data is transmitted from the tag to the reader to determine whether the data has been corrupted or lost. The cyclic redundancy check is one form of checksum.

chip   *See* integrated circuit (IC).

chipless RFID tag   A chipless RFID tag (also known as RF fibers) does not use an integrated circuit to store information. The tag uses fibers or materials that reflect a portion of the reader's signal back; the unique return signal can be used as an identifier. The fibers are shaped as thin threads, fine wires, or labels or laminates. Chipless RFID tags can be used in environments that differ from those with RFID tags with electronic circuitry. They tend to work over a wider temperature range, and they are less sensitive to RF interference. Chipless tags are sometimes used in anti-counterfeiting with documents. However, since the tags cannot transmit a unique serial number, they are less usable in the supply chain.

circularly-polarized antenna   An antenna whose plane of polarization rotates in a circle, making one revolution during the time it takes the radio wave to travel the distance of one wavelength. If the rotation is clockwise, looking in the direction of propagation of wave, the sense is called right-hand-circular (RHC). If the rotation is counterclockwise, the sense is called left-hand-circular (LHC). These antennas are used when the orientation of the tag to the reader cannot be controlled. They provide shorter read range than the linearly polarized antenna because only half the radio energy is transmitted in any one plane at a time.

closed-loop systems   RFID control systems in which the tracked objects never leave the company or organization; the objects either stay within one organization or circulate among a small known group of organizations. All the data related to the object is usually stored in one place, so the complete history of the object is readily available. The usual concerns about RFID standards do not apply, since the tracked object does not leave the system. This type of system is somewhat easier to implement because of the controlled set of variables. Although closed-loop systems may be implemented without concerns for RFID standards, it is a good practice to follow prevailing industry standards in system design.

collaborative planning, forecasting, and replenishment   A general term used to describe cooperation between manufacturers and retailers to match supply of goods with demand for them.

commissioning a tag   The process of writing data to the tag for the first time. The data may include only an identifying number or a number and other information and may be written to a tag during tag manufacturing or when the tag is attached to the object. In most RFID systems, this data is associated with information about the object stored in the database.

compatibility   Two systems are considered compatible if they are equal in interface specifications, such as protocols, frequencies, and voltage levels, and are thus able to operate together. (*See* interoperability.)

concentrator   A device connected to several RFID readers that gathers data from the readers. It performs some filtering and then passes on only useful information from the readers to a host computer.

contactless credit card   Unlike an ordinary credit card, a contactless credit card uses RFID technology to store information about an account and to transfer it to the merchant. Standard credit cards carry data on a magnetic strip; when the card is swiped against a reader, the data is transferred. The primary advantage of contactless cards lies in the speed of the transaction.

coupling   *See* inductive coupling.

cyclic redundancy check (CRC)   A method of checking data stored on an RFID tag to be sure that it hasn't been corrupted or lost. (*See* checksum.)

data carrier   A medium that is capable of holding data, such as RFID tags, barcodes, and magnetic strips.

data field   The smallest subdivision of RFID integrated circuit memory that is used to store a particular type of information, such as a product number or an access password.

data field protection   The ability of an RFID integrated circuit to prevent data stored in a data field from being overwritten.

data transfer rate   The number of bits or bytes of data that can be transferred between an RFID tag and a reader per unit of time.

dB   *See* decibel.

dBm   Unit of power specified in decibel relative to 1 milliwatt.

dead tag   An RFID tag that cannot be read by an RFID reader due to a fault in the tag.

decibel   A unit used to express ratio of two quantities, usually power. Decibel is one tenth of a Bel—a log of ratio of two quantities. Antenna gain and cable losses are specified in decibel.

detune   When RFID tags are attached to a conductive material or are in the vicinity of such material, the property of the antenna on the tag changes. This causes the antenna to be tuned to a different frequency than the intended design frequency. The antenna is said to be detuned from its intended frequency.

die   Silicon disk on which integrated circuits are created.

dielectric   A material that is a poor conductor of electricity, but an efficient supporter of electrostatic fields. Some examples of dielectric include porcelain (ceramic), mica, glass, plastics, oxides of various metals, dry air, and distilled water. Dielectrics are used in RF transmission lines. In a coaxial cable, polyethylene, a dielectric, is used between the center conductor and outside shield.

dielectric constant   Also called relative permittivity. A number that relates the ability of a material to carry alternating current to the ability of a vacuum to carry the same. The dielectric constant of a material affects how electromagnetic signals (light, radio waves) move through the material. A high dielectric constant detunes an RFID antenna and reduces the performance of an RFID tag.

digital certificate   An attachment to an electronic message used for security purposes. It verifies the authenticity of a user sending a message and provides the receiver with the means to encode a reply. An individual wishing to send an encrypted message applies for a digital certificate from a certificate authority, which issues an encrypted certificate containing the applicant's public key and a variety of other identification information.

digital signal processor (DSP)   A specialized microprocessor designed specifically for digital signal processing. It is used in RFID readers.

digital signature   An electronic signature that can be used to authenticate the identity of the sender of a message or the signer of a document. It also ensures that the original content of the message or document that has been sent is unchanged.

dipole   An antenna made of a straight electrical conductor measuring half the wavelength from end to end and connected at the center to an RF feed line. This is the basic and simplest practical antenna and has gain of 2.2 dBi. It is frequently used as a base to compare other antennas. Most UHF and microwave tags have dipole antenna connected to an integrated circuit.

dual interface smart card   An RFID-enabled card that can be read either when it comes in contact with a reader or from a distance.

dumb reader   An RFID reader that does not have a microprocessor and lacks data processing capability.

duplex   A communications system that allows transmission of data in both directions. Full duplex allows transmission of data in both directions simultaneously. Half duplex allows transmission of data in both directions, but not simultaneously.

duty cycle   The percentage of the time the reader is allowed to transmit RF signals within a specified time slot.

EAN   *See* European Article Numbering.

EAS   *See* electronic article surveillance.

edge server   A computer directly controlling data collection devices and peripherals in warehouses, distribution centers, and manufacturing plants. This computer supplies collected data, sometimes after filtering, to enterprise servers.

EDI   *See* electronic data interchange.

EEPROM   *See* electrically erasable, programmable, read-only memory.

effective isotropic radiated power (EIRP)   The amount of power that would have to be emitted by an isotropic antenna to produce the peak power density observed in the direction of maximum antenna gain. EIRP generally takes into account the losses that occur in transmission lines and connectors and the gain of the antenna. EIRP is used by many regulating authorities to specify maximum allowable power from an antenna. It is sometimes referred to as equivalent isotropically radiated power.

effective radiated power (ERP)   The power necessary at the input terminals of a reference half-wave dipole antenna to produce the same maximum field intensity. It is a product of the power supplied to the antenna and antenna gain relative to a half-wave dipole in a given direction, usually in the direction of maximum gain.

EIRP   *See* effective isotropic radiated power.

electrically erasable, programmable, read-only memory (EEPROM)
Read-only memory that can be erased and reprogrammed repeatedly through the application of higher than normal electrical voltage. RFID tags that use

EEPROM are more expensive than factory programmed tags that cannot be re-programmed.

electromagnetic interference (EMI)   Interference caused by radio waves of various RF-emitting devices such as cell phones, cordless phones, and microwaves. EMI reduces the performance of an RFID system.

electronic article surveillance (EAS)   A method of preventing unauthorized removal of articles from an area. The system consists of gate antennas, readers, and EAS tags. Tags are attached to the articles being monitored. This system is widely used to prevent shoplifting in retail stores and pilferage of books from libraries. The tags are turned off by the staff when items are properly removed. If a tagged item is removed without the tag being deactivated, the detection system located at the exit reads the tag and sounds an alarm or alerts the staff.

electronic data interchange (EDI)   A method of transmitting electronic data in a standardized format.

electronic pedigree   Collection of data about the movement of product through supply chain, which helps prevent product counterfeiting.

Electronic Product Code (EPC)   A family of coding schemes designed by EPCglobal, Inc., to provide unique identification of objects in global trade. The object can be a product, a container, a vessel, or a location.

EMI   *See* electromagnetic interference.

encryption   A method of scrambling information to make it unreadable to those who are not authorized to read it. In an RFID system, encryption prevents readers that are not part of the system from reading information stored on RFID tags.

enterprise resource planning (ERP)   A business management multi-module software system that integrates (or attempt to integrate) all facets of the business, including planning, manufacturing, sales, and marketing into a unified system. It integrates business activities such as product planning, purchasing, inventory control, order tracking, customer service, finance, and human resources. All information is saved in a relational database system. The deployment of an ERP system can involve considerable business process analysis, employee retraining, and new work procedures.

EPC   *See* Electronic Product Code.

EPC Discovery Service   As a part of EPCglobal Network, allows users to find data related to a specific Electronic Product Code.

EPC Generation 2 (Gen 2)   The standard ratified by EPCglobal for the air interface protocol for the second-generation of EPC technologies.

EPC Information Service   A specification for a standard interface for accessing Electronic Product Code–related information from EPCglobal Network. It allows companies to store EPC data in secure databases on the Web.

EPCglobal, Inc.   A non-profit organization set up by the Uniform Code Council and EAN International, the two organizations that maintain barcode standards, to commercialize EPC technology. EPCglobal is leading the development of industry-driven standards for EPC to support the use of RFID. Uniform Code Council and EAN International are now part of an organization called GS1.

EPCglobal Network   A set of specifications that allows companies to collect, store, and retrieve data associated with Electronic Product Codes.

EPROM   *See* erasable, programmable, read-only memory.

erasable, programmable, read-only memory (EPROM)   Nonvolatile memory that can be erased by exposing it to ultraviolet light and reprogrammed.

error correcting code   Code stored on an RFID tag that enables localization and correction of errors.

error correcting mode   Automatically detects and corrects errors in the data transmission between tags and readers.

ETSI   *See* European Telecommunications Standards Institute.

European Article Numbering (EAN)   An industry organization that develops and maintains barcoding standards for product identification.

European Telecommunications Standards Institute (ETSI)   An independent, non-profit telecommunications standards organization whose work is carried out by technical working groups consisting of telecommunications companies, manufacturers, regulatory authorities, and other parties in the sector.

excite   An RFID reader excites (powers up) a passive tag by sending radio waves toward it. The tag responds by transmitting a radio signal toward the reader.

eXtensible Markup Language (XML) A general-purpose language used to share information over the Internet.

factory programming The information on some read-only tags is written in the factory where they are manufactured. Such tags cannot be reprogrammed.

false read *See* phantom read.

far-field communication If an RFID tag is outside of one full wavelength of the reader, it is said to be in the far field. The far-field signal decays as the square of the distance from the antenna. The far-field signal is typically used in ultra high frequency (UHF) and microwave systems. *See* also near-field communication.

field programming Writing data to the tag after it has been manufactured. These tags use EEPROM memory that can be erased and reprogrammed many times.

firmware Software that is embedded in a hardware device. A computer program in a read-only memory (ROM) integrated circuit.

Flash memory A form of nonvolatile computer memory that can be electrically erased and reprogrammed in blocks instead of 1 byte at a time.

folded dipole A dipole antenna in which the poles are connected to each other as well as to the integrated circuit.

form factors The different forms in which RFID tags are available, including smart labels, plastic cards, key fobs, and so on.

forward channel The path of transmission of radio waves from the RFID reader to the RFID tag.

frequency The number of occurrences of a repeated event per unit time, measured in Hertz (Hz). RFID systems use low, high, ultra high, and microwave frequencies.

frequency hopping A technique to prevent RF devices operating in the same frequency range from interfering with one another.

frequency shift keying (FSK) A modulation technique involving two different frequencies used in digital signals. The two binary states, logic 0 (low) and 1 (high), are each represented by an analog waveform. Logic 0 is represented by a wave at a specific frequency, and logic 1 is represented by a wave at a different frequency.

FSK *See* frequency shift keying.

GCI *See* Global Commerce Initiative.

GDS *See* global data synchronization.

Gen 2 *See* EPC Generation 2.

GLN *See* global location number.

Global Commerce Initiative (GCI) A user group that brings manufacturers and retailers together on a worldwide parity basis to simplify and enhance global commerce and improve consumer value in the overall retail supply chain.

global data synchronization (GDS) A process of ensuring accuracy of product information among various supply chain partners such as manufacturers and retailers.

global location number (GLN) A 13-digit number used to identify parties and their locations uniquely in electronic commerce transactions and encoded in a barcode or an RFID tag.

global positioning system (GPS) Satellite navigation system that transmits signals allowing GPS receivers to determine the receiver's location, speed, and direction. It is an important tool for map-making and land surveying.

Global System for Mobiles (GSM) A digital mobile telephone system that is widely used in various parts of the world. It operates on the 900-megahertz and 1.8-gigahertz bands in Europe, where it is the predominant cellular system, and on the 1.9-gigahertz band in the United States.

Global Trade Item Number (GTIN) Used for identifying trade items and is developed by GS1. Product identification numbers, such as EAN/UCC-8, UCC-12, EAN/UCC-13, and EAN/UCC-14, are drawn from the worldwide system of GTIN.

GPS *See* global positioning system.

ground plane An electrically conductive surface that may be natural, such as earth or ocean; an available artificial surface, such as vehicle body; or a specially designed artificial surface, such as a large conductive metal plate found in most antennas. The ground plane should extend a minimum of one wavelength (preferably more) in each direction from the antenna.

GS1   GS1 is a leading global organization dedicated to the design and implementation of global standards and solutions to improve the efficiency and visibility of supply and demand chains globally and across sectors.

GSM   *See* Global System for Mobiles.

GTIN   *See* Global Trade Item Number.

harvesting   A technique in which energy from an RFID reader is gathered by a passive RFID tag.

high frequency (HF)   A frequency range between 3 and 30 MHz. In RFID systems, only 13.56 MHz is used out of the entire HF range.

host system   A networked computer that provides services to other computers or users in that network.

hybrid card   A card with two chips—one with a contact interface and one with a contactless interface; unlike a dual-interface card, which has a single chip with both contact and contactless interfaces.

IC   *See* integrated circuit.

inductive coupling   The transfer of energy from one circuit to another by means of mutual inductance between the two circuits. Some RFID tags and readers exchange information using inductive coupling between their antennas. Low-frequency and high-frequency RFID systems use inductive coupling.

Industrial, Scientific, and Medical (ISM) bands   Radio frequency ranges (902–928, 2400–2483.5, 5725–5850 MHz) originally reserved internationally for non-commercial use by the Industrial, Scientific, and Medical (ISM) organizations. In recent years they have also been shared with license-free error-tolerant communications for commercial applications such as wireless LANs and Bluetooth.

inlay   Also known as inlet, it consists of an integrated circuit, an antenna, and a substrate. It is the main component of an RFID tag and smart label.

inlet   *See* inlay.

input/output ports (I/O ports)   Ports on an RFID reader that can be connected to external devices.

integrated circuit (IC)   A small electronic circuit that is created on the surface of a thin substrate of a semiconductor.

intelligent reader   A reader that incorporates an embedded computer system that can filter the data collected from the RFID tags, control peripheral devices, and execute commands, unlike a dumb reader that can only retrieve data from RFID tags.

intentional radiator   A device that produces radio waves on purpose, including RFID transmitters.

interconnection   A device that connects an RFID integrated circuit to an antenna on an RFID tag.

International Organization for Standardization (ISO)   A non-governmental organization made up of the national standards institutes of 146 countries that produces world-wide industrial and commercial standards.

interoperability   The ability of RFID tags and readers from different vendors to operate in an RFID system.

interrogation zone (IZ)   The space in which an RFID tag can receive a radio signal from a reader. For a passive tag the signal should be strong enough to power the tag, while for a semi-passive tag it should be strong enough to raise the tag's backscattered signal above the noise floor.

interrogator   *See* reader.

I/O ports   *See* input/output ports.

ISM   *See* Industrial, Scientific, and Medical bands.

ISO   *See* International Organization for Standardization.

ISO 7816   An international standard for smart cards.

ISO 10536   An international standard for contactless integrated circuit cards.

ISO 11784 and 11785   International standards that regulate the radio frequency identification of animals in regards to code structure and technical concept.

ISO 14443   Defines international standards for a proximity card used for identification that usually uses the standard credit card form factor.

ISO 15693   Defines international standards for vicinity cards.

ISO 18000   A series of RFID air interface standards for the item identification.

isotropic antenna   An antenna that emits energy equally in all directions. A theoretical concept for which no physical antenna can be made.

item-level tagging   Using RFID tags to track individual items throughout the supply chain.

linearly-polarized antenna   An antenna that emits energy from the reader wholly in one plane containing the direction of propagation.

low frequency (LF)   Low frequencies range from 30 to 300 kHz. Low-frequency RFID systems usually operate at only one of two frequencies—125 kHz or 134 kHz.

memory block   Data stored on an RFID integrated circuit is divided into different sections known as memory blocks. Each memory block can be individually read from and written to, and can be locked to prevent it from being overwritten.

memory capacity   The amount of data that can be stored on an RFID integrated circuit.

microcontroller   A tiny computer on a chip. It is a single integrated circuit comprising a central processing unit, input/output interfaces such as serial ports, peripherals, RAM for data storage, and ROM, EPROM, EEPROM, or Flash memory for program storage.

microprocessor   A single integrated circuit that contains a CPU and some form of memory.

microwaves   Electromagnetic waves that range between 1 GHz and 300 GHz frequencies. RFID systems use only two frequencies—2.4 and 5.8 GHz.

middleware   In an RFID system, a set of software modules that collects data from RFID readers, controls and monitors RFID readers and peripherals, and filters information received from RFID readers. The information is then passed to the enterprise applications. Middleware acts as a conduit between the edge devices and business applications such as factory and warehouse management systems.

modulation   The process of modifying the carrier wave to transmit information. Various types include amplitude modulation, phase modulation, frequency modulation, and pulse-width modulation.

monostatic reader   An RFID reader that uses the same antenna to transmit and receive radio signals.

multiplexer   Allows several single antennas and gate solutions to be operated with only one reader. A device that allows several antennas to be connected to a single antenna port. Multiplexers are not efficient and economical to use at higher frequency ranges such as UHF and microwave.

National Institute of Standards and Technology (NIST)   A federal technology agency whose mission is to promote U.S. innovation and industrial competitiveness by advancing measurement science, standards, and technology in ways that enhance economic security and improve quality of life.

near-field communication (NFC)   The tag and reader antenna are coupled within one full wavelength of the carrier wave. The near-field signal attenuation is proportional to the sixth power of the distance from the antenna.

NFC   *See* near-field communication.

nominal range   The range at which a tag can be read reliably.

non-volatile memory   Computer memory that can retain stored information even when the power to it is turned off, such as EPROM, EEPROM, and Flash.

null spot   A space in the interrogation zone that does not receive a strong enough radio signal to promote communication between a tag and a reader.

Object Naming Service (ONS)   A subsystem within the EPCglobal Network that provides the location of an EPC Information Service for an Electronic Product Code number. It operates just like a DNS system for Internet access.

OEM   *See* original equipment manufacturer.

one-time programmable　Memory that can be programmed only once; then it becomes read-only memory.

one-time programmable tag　An RFID tag that can be written to only once and read many times.

ONS　*See* Object Naming Service.

orientation　Alignment of the RFID tag with respect to the reader antenna polarization plane. To communicate with a reader connected to a linearly polarized antenna, a tag must have the same orientation as the polarization plane of the reader antenna. If the reader antenna is circularly polarized, the tag orientation has no effect.

original equipment manufacturer (OEM)　An organization that makes products for others to repackage and sell. Resellers buy OEM products in bulk and then sell them under the reseller brand.

passive tag　An RFID tag that does not contain an on-board power source (a battery) and an active transmitter. It gets its power from radio waves emitted by a reader.

patch antenna　An antenna that consists of a square conductor mounted over a ground plane.

phantom read　The phenomenon in which a reader reads a tag that does not belong to the business process being monitored.

phase shift keying (PSK)　A method of digital communication in which the phase of a carrier wave is changed or modulated to convey information.

Physical Markup Language (PML)　A markup language based on XML for communicating a description of products.

PML　*See* Physical Markup Language.

power level　The amount of energy emitted by an RFID reader, usually expressed in watts or dBm.

private key　Encryption/decryption key known only to the party or parties that exchange secret messages.

programming a tag   Encoding information on an RFID integrated circuit.

protocol   *See* air interface protocol.

proximity sensor   A device that detects the presence of an object. Proximity sensors may be of contact or non-contact type.

PSK   *See* phase shift keying.

public key   A publicly available key provided by some designated authority as an encryption key that, combined with a private key, can be used to encrypt messages and digital signatures.

quiet tag   An RFID tag that can be read occasionally with the interrogator output at full power, or that can be read only at very close range.

Radio Frequency Identification (RFID)   A method of identifying objects using radio waves. Typically, a reader communicates with a tag, which holds digital information in an integrated circuit. Chipless forms of RFID tags use material to reflect back a portion of the radio waves beamed at them.

RAM   *See* random access memory.

random access memory (RAM)   Memory used for temporary storage of data. Information stored in RAM is lost when power to it is removed.

read   The process of retrieving data stored on an RFID tag.

read range   The maximum distance in which a reader and tag can communicate.

read rate   The number of tags that can be read per unit of time.

read-only tag   An RFID tag that can only be read. The data written on its integrated circuit cannot be changed.

read/write tag   An RFID tag that can be written to and read from many times.

reader   A device that retrieves information written on RFID tags by transmitting and receiving radio signals to and from the tags. Reader also provides power to passive tags, and most readers can also write data to the tag.

reader field   A space within which an RFID tag can be read by the reader.

reader talks first   A design in which a reader initiates the communication with the tag.

real time location system (RTLS)   A system for tracking the location and movement of products in real time using RFID tags and readers.

reverse channel   The channel through which a radio signal travels from an RFID tag to a reader. It is also called return link.

RFID   *See* Radio Frequency Identification.

RFID tag   An electronic device that contains an antenna connected to an integrated circuit. It stores data and communicates that data with an RFID reader using radio frequency waves.

RTLS   *See* real time location system.

SAW   *See* Surface Acoustic Wave.

semi-passive tags   An RFID tag that has on-board power (battery) but no active transmitter. The battery is used to power an integrated circuit and optional environmental sensor. Some semi-passive tags sleep until they are awakened by a signal from the reader, which conserves battery life. These tags are sometimes called battery assisted tags or battery assisted passive tags.

sensor   A device that produces an electronic signal in response to something in the environment. Sensors are increasingly being combined with RFID tags to detect the presence of a stimulus at an identifiable location.

signal attenuation   A weakening of a signal as it travels from a transmitter to receiver.

signaling technique   A complete description of the modulation, encoding, protocol, and sequences required to communicate between two elements of a system.

silent commerce   This term covers all business solutions enabled by tagging, tracking, sensing, and other technologies, including RFID, that make everyday objects intelligent and interactive. When combined with continuous and pervasive Internet connectivity, they form a new infrastructure that enables companies to collect data and deliver services without human interaction. This type of commerce is "silent" because objects communicate and commerce takes place without human interaction.

singulation   A method of identifying one tag from a group of tags present in the reader's interrogation zone.

skimming   Reading an RFID tag on a person without his or her knowledge, or reading a tag surreptitiously.

slap and ship   Placing an RFID tag on a case or pallet just before it is shipped from a supplier. Generally used to meet a customer's mandate. It is also called tag and ship.

smart card   Also called chip card or integrated circuit(s) card (ICC). Any pocket-sized card with embedded integrated circuits. When the card uses RFID technology to send and receive data, it is called a contactless smart card.

smart label   A label that usually contains both a traditional barcode and an RFID tag. As barcodes are printed on smart labels, information is also encoded into the RFID tag by the printer.

strap   A tag subassembly where the integrated circuit (IC) is attached to two large, relative to IC, metallic patches by the IC manufacturer. The strap makes it easy to attach antenna to IC during tag manufacturing.

Surface Acoustic Wave (SAW)   An acoustic wave traveling along the surface of a material having some elasticity, with an amplitude that typically decays exponentially with the depth of the substrate. This technology is used for automatic identification in which low power microwave radio frequency signals are converted to ultrasonic acoustic signals by a piezoelectric crystalline material in the transponder. Variations in phase shifts on the reflected signal can be used to provide a unique identity.

synchronization   Allocating time slots to the readers located in close vicinity so they do not interfere with each other.

tag   *See* RFID tag.

tag excitation device (TED)   A term coined by the RFID Alliance Lab for the device that sends signals to the tag regardless of the make or manufacturer. TED is used to measure the response of tags scientifically.

tag talks first (TTF)   A protocol for exchanges between the reader and the tag, where the tag sends information continuously, without waiting for a specific command from the reader.

track and trace   The process of storing and retrieving information about the movement and location of objects.

transceiver   A device that has a transmitter and a receiver; both are combined and share common circuitry or a single housing.

transponder   *See* RFID tag.

ultra high frequency (UHF)   Ultra high frequencies range between 300 and 3000 kHz. UHF RFID systems operate within a narrow range of 860 to 960 MHz.

Uniform Code Council (UCC)   The nonprofit organization that overseas the Uniform Product Code, the barcode standards used in North America.

Unique Identifier (UID)   A globally unique identifying number used by the U.S. Department of Defense to track objects within its system.

Universal Product Code (UPC)   The barcode standard used in North America and administered by the Uniform Code Council.

WORM   Write once read many. A tag that can be written once by the user and thereafter can only be read.

write   The process of transferring data to a transponder from the reader.

write rate   The rate at which information is transferred, written into the tag's memory, and verified as being correct.

XML   *See* eXtensible Markup Language.

Index

A

AB symmetry, 116
Access control points, 196-197
Active antenna arrays, 119
Active tag frequencies, 76
Active tag periodic beacon, 76
Active tag read range, 158
Active tag sleep mode, 76
Active tag specifications, 95-96
Active tags, 75-76, 158
AIDC (Automatic Identification and Data Capture), 248
Air interface, 85
Air interface protocols, 112-113
Air ionizer, 171-172
ALE (Application Level Events) standard, 8-9
Alerts, 142-143, 170
Alien Technologies semi-passive tag specs, 94-95
AM (amplitude modulation), 30
Amplitude triangulation system, 145-148
Amplitude of a wave, 17, 29, 145-148
Anechoic foam, 171
Animal tracking, 160
Antenna angling in a portal, 191-192
Antenna aperture, 45
Antenna arrays, 52-53
Antenna beam width, 45, 185-186
Antenna beams, 45, 53, 167, 185-186
Antenna cable, 55-56, 169, 184, 187
Antenna datasheets, 185-186
Antenna design, 41-51, 64-65
Antenna footprint, 111
Antenna gain, 28, 45-50, 166-167
Antenna installation, 185-187
Antenna orientation, 209
Antenna performance, 50-51
Antenna polarization, 48-50, 89, 111-112, 166
Antenna positioning, 119
Antenna radiation plot, 111
Antenna selection, 166-168
Antenna settings, 117
Antenna side lobes, 46
Antenna size, 168
Antenna types, 51-55
Antenna whitepapers, 185-186
Antennas, 3, 64-65, 110-112
    bandwidth of, 43
    cabling to reader, 169, 184

in dock stands, 168
environment of installed, 167-168
far-field of, 31-32
IC attachment, 65-67
IC connection, 65
impedance of, 44-45
mounting, 168, 185-187
near-field of, 31-34
overlapping coverage from, 111
performance and characteristics of, 41-57
receive, 41, 186
resonant frequency of, 42-43
ruggedness of, 167-168
theoretical, 51
transmit, 41, 186
Anti-collision methods, 116
Antistatic bag, 172
Applications for RFID systems, 9-11, 251-252
ASK (amplitude shift keying), 30
ASN (advance ship notice), 11
Asset tracking, HF tags for, 79
Australia, regulations in, 269-270
Auto-ID (automatic ID) technologies, 2-4
Automatic label applicators, 138-140
Automation, RFID, 9
Automotive industry
    ignition circuits, 160
    LF tags embedded in car keys, 78

B

Backscatter (radiative) coupling, 30-32, 34-36
Baggage tracking, HF tags for, 79
Bandwidth of an antenna, 43
BAP (battery-assisted passive) tags, 73
Barcode print methods, 131
Barcode scanners, 3
Barcodes, 2
    advantages and disadvantages of, 4
    history of, 4-5
    versus RFID technologies, 3-5
BATs (battery-assisted tags), 73
Batteries, for semi-passive tags, 73-74
Beam antennas, 45, 53, 167, 185-186
Beamwidth of an antenna, 45, 185-186
Bel, defined, 28
Biometrics, 2
Bistatic interrogator, 107, 186
Blow label applicators, 139

305

Empowering U through knowledge

- ➡ **Classroom Training In RFID Labs**
- ➡ **Online Training**
- ➡ **eLearning**
- ➡ **Advisory Services**

RFID4U™ is a global provider of RFID Knowledge, Education and Advisory Services with projects throughout the US, Europe and Asia. Our experienced consultants are industry recognized and trusted Subject Matter Experts. They are known for their participation in major industry initiatives, authoring respected white papers and books and speaking at major trade shows and industry events. RFID4U partners with the best RFID manufacturers, service providers and laboratories throughout the world. We use cutting edge technologies to meet business challenges in all industry verticals throughout the world.

RFID4U
Corporate and International Headquarters
355 W. Olive Avenue, Suite 207
Sunnyvale, CA 94086

Phone: (408) 739-3500
Toll-free: (866) 400-RFID
Fax: (408)739-3502
Email: info@rfid4u.com
Web: www.rfid4u.com